Rethinking the Environmental Impacts of Renewable Energy

Renewable energy is important as a substitute for finite fossil fuels and inflexible nuclear power and could conceivably power the world. However, this is challenging as the world is currently 80 per cent dependent on fossil fuels, and renewable sources produce only about 15 per cent of total energy. Conversion technologies for use with many of the eight different primary sources of renewable energy are only just emerging as viable technologies. While renewable energy sources will not run out, and their use involves little or no release of carbon dioxide or ionising wastes, they do have local environmental impacts of their own.

This book analyses the nature of environmental impacts from renewable sources. A novel method of assessing impacts is explored based on a set of parameters centred on how diffuse or concentrated the energy flow is. The approach that is developed will inform engineers, designers, policy makers and planners, as well as researchers in the area.

Alexander Clarke is a Teaching Fellow in Renewable Energy in the Natural Sciences Department at the University of Bath, UK, and an Associate Lecturer at the Open University.

Rethinking the Environmental Impacts of Renewable Energy

Mitigation and management

Alexander Clarke

LONDON AND NEW YORK

First published 2016
by Routledge
2 Park Square, Milton Park, Abingdon, Oxon OX14 4RN

and by Routledge
711 Third Avenue, New York, NY 10017

Routledge is an imprint of the Taylor & Francis Group, an informa business

British Library Cataloguing in Publication Data
A catalogue record for this book is available from the British Library

Library of Congress Cataloging-in-Publication Data
Names: Clarke, Alexander, author.
Title: Rethinking the environmental impacts of renewable energy: mitigation and management/Alexander Clarke.
Description: Abingdon, Oxon; New York, NY: Routledge, 2016. |
Includes bibliographical references and index.
Identifiers: LCCN 2015049366 | ISBN 9780415722179 (hardback) |
ISBN 9780415722186 (pbk.) | ISBN 9781315616605 (ebook)
Subjects: LCSH: Renewable energy sources—Environmental aspects.
Classification: LCC TD195.E49 C548 2016 | DDC 333.79/414—dc23
LC record available at http://lccn.loc.gov/2015049366

ISBN: 978-0-415-72217-9 (hbk)
ISBN: 978-0-415-72218-6 (pbk)
ISBN: 978-1-315-61660-5 (ebk)

Typeset in Bembo
by Keystroke, Station Road, Codsall, Wolverhampton

Contents

Illustrations

Figures

Tables

Foreword

All forms of energy use incur environmental impacts, and although renewable energy seeks to avoid some of the more intractable ones such as global warming or ionising radiation and long-lived wastes, there are still impacts associated with it.

This book is written in the full knowledge that the topic of environmental impacts of renewable energy can be a contentious subject which has become, in some countries, something of a battleground over the type of energy supply. However, the aim of this work is not to supply ammunition to one side or the other, but to constructively address some of the real dilemmas and issues at the heart of utilisation of natural energy flows. This is attempted by seeking to make more sense of the subject, through a rigorous analysis based on scientific principles.

Some may choose to use this material to try to show that renewable energy is impractical, damaging, or unaffordable. That, however, is not the position of the author. Nevertheless, the topic is one of great complexity, which should be amenable to general principles, but also subject to particular local conditions. Renewable energy may be a 'good thing' per se, as it is unending and largely free from greenhouse gas emissions, though with the proviso that there is a balance of interests and also that it is implemented in an optimal manner.

Most people will be aware that there are powerful vested interests in the energy world, and these are the 'incumbents' which renewable energy sources aim to displace. As might be expected, the incumbent energy interests will not necessarily lightly give up their role; resistance can be expected. This might, for example, be in the form of misinformation. Various myths can be promulgated, e.g. that renewable sources are too variable, intermittent and diffuse, or have too many environmental impacts to be practical or economic. Examples of the successful deployment of renewable energy could be deliberately ignored in order to strengthen the argument. For instance, those countries that produce a significant proportion of their energy from renewables could be left out of the debate, or portrayed as misguided. The prevailing discourse in mainstream media could refer to renewable energy sources in a somewhat pejorative manner, with a continuous drip-feed of environmental problems, rather than informing the public of its successes.

In general, awareness of energy matters is not great and there are real issues to confront in deploying large-scale renewable energy supply. It is perhaps inevitable that there could be misinterpretation of the parameters of impact investigated here. As stated below, since environmental impact does indeed constitute a constraint on renewable energy, it may be futile to hope that will not happen.

Looking ahead to a time when it will no longer be acceptable to use fossil fuels – i.e. the sequestered biomass carbon – which natural systems put away as it were *for a purpose*, as energy sources, we will need to rely on renewable energy or atomic power. Coming back to the present, the temptation to use all the remaining fossil fuel in the ground seems too great – it is too easy to carry on with business as usual. Renewable energy sources are in many respects more complex and subtle than merely burning 'old sequestered biomass'.

Historically, the assumption has been that energy sources and flows are incidental, i.e. they serve no function in nature and therefore are readily available for human use, whether natural sequestered biomass (fossil fuels) or natural renewable energy flows. This book questions that approach and tries to identify and assess the role of energy flows in the natural environment.

Acknowledgements

I would like to acknowledge the advice and help given by several people in putting this book and study together. In particular I am grateful to Professor David Elliot for his advice and editorial assistance and Dr Chris Whitlow for his help and advice in constructing the test. Thanks are due to several others who have had an input to the study, including Professor Godfrey Boyle and Professor David Coley.

While efforts have been made to ensure the accuracy and validity of statements and calculations in this book, the breadth of the topic has in places necessitated the use of somewhat coarse techniques. Any resulting errors and mistakes are all my own.

1 Introduction

1.1 Introduction

Renewable energy is important as a substitute for finite and polluting fossil fuels and inflexible and ionising nuclear energy. It could conceivably power all of humanity's energy needs, since the sun sends sufficient energy to earth to power human energy use many thousands of times – a huge resource. There could be many advantages to relying on renewable energy, in terms of stability of supply, environmental effects and cost, to name just a few. In fact, the world already does rely on renewable energy to an extent, as about 13–15 per cent or more of primary energy is derived from renewable sources – and always has been. This is mostly biomass (the use of plant and animal matter for energy) at about 10 per cent and hydro electricity at about 3 per cent primary energy but producing about 16 per cent of the world's electricity. These sources can be considered as the traditional renewable energy sources, used for thousands (maybe 50,000 or more) of years in the case of biomass and for about 120 years in the case of hydro electricity. However, there are a variety of newer renewable energy sources, such as solar, wind, wave, tidal and geothermal which are now emerging as viable energy providers. Some of these, such as wind and solar, are now expanding rapidly, at up to about 20–30 per cent per year or more, while others are still relatively immature technologies but showing promise. As new renewable energy sources and technologies emerge, it is increasingly feasible to look ahead to a time when renewables will be able to provide energy for all of humanity's needs. That would require a 6-fold increase in renewable supply to provide for today's energy needs. However, much of humanity is still effectively starved of access to adequate energy, and given population growth it is more likely that at least a 20-fold increase in renewable energy supply will be required. Is this still feasible? The answer is yes, as the resource is thousands of times greater than human energy use, as stated above. However, although the solar resource may be thousands of times that of human energy use, other sources are not so abundant. Moreover, renewable energy sources are not incidental; they power the natural environment. The fundamental processes of the earth, such as the maintenance of hospitable conditions for life, the warmth of the atmosphere and oceans, the hydrological cycle and photosynthesis, are all dependent on these natural energy flows. If a 20-fold increase in renewable energy supply is wanted, it will be important to understand how to achieve this without adversely affecting these vital natural systems. In fact, one of the main advantages of renewable sources over fossil fuel should be that it incurs less perturbation of natural systems. So it will be necessary to design and choose our resource exploitation carefully in order to avoid adverse effects and depletion.

The scale of the challenge is great, since the world is roughly 80 per cent dependent on fossil fuels and many of the renewable energy sources have only just emerged. At least eight different primary renewable energy sources exist: solar, biomass, hydro, wind, tidal, marine current, wave and geothermal. There are a plethora of different technologies available to convert these sources into useful energy forms. All of the great variety of sources will need to be hugely scaled up and developed to provide viable replacements for conventional energy sources. While renewable energy sources are essentially benign, in that, unlike fossil fuels, they do not contribute to irreversible global warming and climate change or result in long-lived ionising wastes, they do have their own local environmental impacts. Attention needs to be paid to such impacts, as cumulatively they constitute a real constraint on renewable energy capacity. Moreover, as renewable energy is deployed at scale, for example offshore wind farms of 600 MW such as the London Array, or the roll-out of solar farms occupying several square kilometres, more fundamental environmental issues of abstraction of natural energy flows come into play. Such natural energy flows power the natural environment, in all its myriad forms, from climate to wildlife and the food source for humanity. If we are to intercept and abstract significant quantities of energy from such flows it will be vital to understand what we are doing, and importantly what the possibilities, limitations and constraints are. What are the factors that determine whether the impacts of substantial extraction of energy are significant? How can such impacts be minimised? How should we tap into such sources? This book is not and cannot be a comprehensive description and account of all environmental impacts from renewable energy sources. Instead it investigates an approach to mitigation and management of those impacts. This book explores the parameters and dimensions of factors likely to determine impact, starting with the 'power flux density' or intensity of natural energy flows and what this means in practice to such impacts as land use or interruption to functions in nature.

Although in any one area there are usually several different sources available for exploitation, it is generally agreed that the ultimate limitations to the deployment of renewable energy will be the size of the resource and environmental constraints. In the author's view, environmental impacts of renewable energy are currently considered in a somewhat ad-hoc manner and are not well characterised. There appears to be only patchy understanding of the fundamental principles determining environmental impact.

As renewable energy exploitation rapidly expands, it will be important to understand which factors are the prime limiting and shaping ones. Identification of the key factors involved would provide policy makers, engineers and technologists with suitable tools for decision making and design choices.

While there are considerable advantages to using renewable energy, nevertheless exploitation of renewable energy will involve tapping into natural energy flows, altering them and effectively competing with their other functions, both in nature and those related to human activities. For example, extracting energy from rivers can reduce sediment transport, and desert solar power stations can effectively sterilise the land below them. The use of biomass energy can directly compete with food production, as the recent debates over food versus fuel can testify. On a scale of the hierarchy of need, food production for a still-growing world population could outweigh energy production, all things being equal. Wildlife in natural rivers modified for energy extraction, or rainforests converted to palm oil plantations, can then suffer the effects. Therefore it is important to discover how best to tap into renewable energy resources, for example how much energy can safely be extracted from a flow, and exactly how we should be extracting it so as to avoid undue impacts.

As such issues are being confronted, one of the great benefits of modern technology design is that there is an increasing variety of alternative technical options available. More than ever, schemes can be designed and tailored to fit appropriate objectives. A wealth of environmental science and data now exists, so that lack of knowledge can no longer be an excuse for inappropriate developments (Petts 1984). This process of adjustment to availability of data and wider design choices will involve much effort in reconciling the different disciplines and bringing their considerations and information to bear on energy developments that will be badly needed (Holdren *et al.* 1980). In the rush to develop new energy resources it will be necessary to ensure environmental objectives are not lost to energy objectives.

In seeking to find solutions to the apparently contradictory demands on these resources (maximum energy output, minimum environmental impact), new concepts are required so that human ingenuity can be brought to bear. In this book the emphasis is on the parameter 'power flux density', and changes to it, as a key factor. As a measure of how intense or diffuse the energy flow is, power flux density is defined as kilowatts per square metre in this book. It is proposed that this may be a key factor in the environmental impact of renewable energy sources, and one of the main aims here is to examine this assertion.

1.1.1 Nature of the study

The study considers mainly the physical direct effects on the environment of renewable energy, from capturing, abstracting and converting some of the energy from the natural energy flows such as sunlight, wind and flowing water. The study does not in most instances address the 'upstream' impacts, such as the carbon emissions from producing the materials and manufacturing involved in renewable energy conversion devices and structures. It is considered that much work has already been carried out on this issue. Such impacts can in some cases be significant, especially where the particular renewable energy technology is 'materials intensive' (Mortimer 1991), i.e. where the converter devices and structures involve a large mass of materials per kilowatt of energy captured, or where noxious materials are used. An example of the latter might be the use of gallium arsenide in some solar photovoltaic cells. However, the approach of this study is that a 'materials intensive' renewable energy conversion technology may also represent an immature technology, as yet lacking optimally engineered materials use. In terms of this study, the power (kW) captured per unit of materials mass can be related to the energy flux density; and the 'upstream' effects can then be deduced using existing data. In contrast to studies of the external costs of using energy, e.g. the ExternE study (ExternE 1995), this study does not have the economic costs as its prime focus.

The study, as stated, takes a different approach from that of previous studies in that it looks explicitly at energy *flows* rather than assuming that energy equates to fuels or is *store*-based, an example of which is the 'Fuel Cycle' approach of the ExternE study. The rationale for this is that previous studies have been dominated by fossil fuel and nuclear technology concerns, such as the Fuel Cycle, particulates emissions or nuclear wastes. Such concepts are not appropriate to most of the renewable energy sources, with the exception of biomass.

Use of a Fuel Cycle approach is understandable given the current predominance of fossil and nuclear sources in world energy use. However, this study looks ahead to a time when greater renewable energy capacity will be demanded by policy makers, and the status of renewable energy will be elevated. This study considers eight different renewable energy sources, but focuses largely on hydro electric power for its investigation on the basis that this is the oldest of the modern renewable energy sources, and therefore there are considerable data available.

1.1.2 Outline of the book

Having outlined the basic rationale for this study, and the structure of the book, the remainder of Chapter 1 introduces some key definitions and develops some initial analysis concerning factors that influence environmental impacts. Chapter 2 then reviews the literature on environmental impacts from the use of renewable energy sources. In Chapter 3 some basic theory is presented and the hypothesis is explained, while Chapter 4 describes the questions it raises and the approach adopted to setting up a test. In Chapter 5, the setting up of the test is described, along with the cases and data sources involved, and continues with the results from the test, analysis and synthesis. The first conclusions from applying the hypothesis to hydro electric power are drawn. Chapter 7 discusses an extension of the hypothesis, on a qualitative basis, to other water-based renewable energy sources. Chapter 8 extends the hypothesis further to the low energy flux density sources such as wind and solar. Chapter 9 discusses the overall conclusions.

1.2 Definitions

1.2.1 Definition of the environment

The term 'environment' is defined in systems theory as that which is outside the boundary of the system, itself consisting of interconnected parts (von Bertalanffy 1968; Boulding 1956). Bertalanffy distinguishes between closed systems and open systems, where energy and materials are exchanged across the system boundary. In more general usage, environment can mean surroundings, or conditions, the spatial context and the outside links, biological and physical, to the wider world (Blowers and Smith 2003). It includes processes and systems, for example life-support systems. The concept involves boundaries – where these are drawn, as well as issues of scale both in time and space.

However, the environment is not a single entity – it can be divided into at least three basic components for the purposes of impact assessment.

1 the natural living environment;
2 the natural inanimate environment;
3 the human or cultural environment.

(Clarke 1993)

The natural living environment

This is comprised of flora and fauna as it exists independent of human activity. It requires energy to fuel its growth and maintenance, in addition to the maintenance of climatic, soil and water conditions within a certain range. The activities of flora and fauna help to create and influence these conditions.

The natural inanimate environment

This consists of the earth's surface, the oceans and the atmosphere, which are acted upon by energetic processes. The action of energy on the earth's surface is known as geomorphology, and on oceans and the atmosphere oceanography and meteorology, respectively. These processes are an important factor in the morphology of landscapes, e.g. the hills and valleys, and also the maintenance of soil and aquatic conditions. Energy flows contribute to the shape of shorelines, beaches and marine conditions. They can be important to the continuation of existing shorelines and other features.

The human or cultural environment

This category includes, first, human activities on the environment to cultivate wild areas, changing them into, for example, agricultural land, forestry plantations or urban environments. Managed landscapes are achieved through such means as drainage works, earth moving or changing the vegetation.

Second, the human need for certain minimum environmental standards of air and water quality and noise and light requirements, is included here. Third, under this category come human aesthetic notions about the environment, such as the value of landscapes or certain features in the landscape.

The human or cultural environment concerns humans' interactions with the other two parts of the environment, and the degree to which humans interfere with those categories, ranging from high interference – e.g. intensively farmed areas or cities – to low – e.g. traditional fisheries on largely unregulated wild rivers.

Clearly the animate and inanimate categories of the natural environment are of great importance to the human environment. While humans are merely another form of life, human demands on the environment have changed it enormously. The inanimate environment has a determining effect on the patterns of human and other living organisms' activity through climate and geological processes and it might be considered to be a more fundamental category than the other two. But both humans and other living organisms have, in their turn, the capability to influence events in the inanimate environment. Though the scale of geomorphic or meteorological events is very large, so too can be the impact of humans in desertification, or on large areas of tropical forest. In both cases the local climate can be altered by changes in water retention and evapo-transpiration.

These three categories, the natural living environment, the natural inanimate environment and the human or cultural environment are not always tidy ones, but are grouped this way as each can be considered to have its own requirements or functions, or even 'purposes'.

1.2.2 Environmental impact/effect

The term environmental impact has been defined and used in a variety of different ways. Catlow and Thirlwall (1976) draw a distinction between environmental *effects* and environmental *impacts* in a UK Department of the Environment report on environmental impact analysis. They state that while effects are the physical and natural changes resulting directly or indirectly from development, the impacts are the consequences or end products of those effects, represented by aspects of the environment on which we can place an objective or subjective value. The way in which the term is employed in this book is not quite the same, for although distinguishing between effects and impacts appears to convey precision, in practice it is hard to decide where to draw the boundary line.

Another term that may be usefully related to environmental impact is 'environmental stress'. This has been described as 'any force that pushes the functioning of a system or subsystem beyond its ability to restore its former structure and function' (Meir 1972, cited in Petts 1984). This is useful as it can describe a mathematical process. In more modern parlance this can be described as the 'tipping point'.

In this book the term 'environmental impact' is used in the sense of changes to prevailing environmental conditions and is close in meaning to environmental 'effects'. Both of these terms should be amenable to some form of measurement. In describing environmental impacts as changes, these could comprise those that have a greater significance and permanence than mere 'effects'. However the distinction is hard to make. Impacts, as used here, are not in themselves necessarily either detrimental or beneficial. However, the uncertainty about the effect of these changes leads to a presumption in favour of the status quo, and impacts are usually viewed with suspicion.

The degree, rate and frequency of change and its permanence are important variables when measuring environmental impact. The significance of the changed conditions can be assessed by considering their role in particular environments and ecological systems and will need to be discovered. Any conditions that are critical to the functioning of ecosystems or whole environments will need to be identified.

1.2.3 The role of energy in the environment

Natural energy flows are of fundamental importance to the environment since they provide the energy for the growth and maintenance of living organisms. In addition, the geomorphologic processes that occur in the inanimate environment are the result of the action of energy on the earth, seas and atmosphere. Many of these processes are understood well and have been thoroughly documented, though within particular disciplines.

For any natural energy flow, a set of roles or functions in the environment can be ascribed. For a general description of these functions, see below and Table 1.1.

Energy flows may, in some instances, be critical for functions on which many important processes of the existing environment depend. For instance, river sediments are important for the replenishment of flood meadows or deltas, while the transport downstream of biomass by a river constitutes nutrient and food sources for the flora and fauna. Abstraction of energy from natural energy flows by renewable energy technology may significantly influence such functions with resulting significant environmental impacts. An example of this is sedimentation in the Aswan High dam on the Nile in Egypt (El-Moattassem 1994).

Table 1.1 Some functions of energy flows in the environment

Solar	Primary energy flow to earth, heating of seas, heating of air mass, heating of land mass, evaporation of water – hydrological cycle, photosynthesis.
Wind	Heat transport by air movement, mixing and by water vapour transport and direct movement of water, dust transport, erosion, flora and fauna transport, seed and pollen transport, insect transport.
Biomass	Photosynthesis: primary energy (first trophic level) for all living matter, support and growth of biosphere, breakdown of rock and soil formation, micro-climatic effects.
Hydro	Erosion, sediment transport, deposition of sediment, oxygenation, nutrient and organism transport, flushing.
Tidal	Sediment transport, erosion, nutrient and organism transport, oxygenation, water mixing.
Wave	Erosion of rocks and shoreline, long shore currents transport and deposition of sediment, oxygenation, water mixing.
Ocean thermal and currents	Major energy reservoir/buffer, major heat transport, surface atmospheric effects, ocean nutrient and organism transport.
Geothermal	Mineral conveyance/replenishment, chemical reactions, crust mechanics functions, warm marine upwelling effects on currents and nutrient transport.

Source: adapted from Clarke (1993).

Renewable energy definition

'Renewable Energy sources are those derived from continuous energy flows found naturally in the environment' (Twidell and Weir 1986). There are at least eight different primary renewable energy sources: solar, biomass, hydro, wind, wave, tidal, marine currents, geothermal (Boyle 1996).

Natural energy flows

Interactions occur between energy and the environment – in both directions. The natural energy flows that renewable energy harnesses are very diverse, making comparison between them difficult. The basic types are: solar radiation, solar radiation converted into chemical biomass form, moving air, moving water, ocean thermal temperature differences, heat from geothermal sources.

1.2.4 Environmental effects and impacts of renewable energy

Impacts from using renewable energy can be divided into common effects and source-specific effects, as has been noted by the author (Clarke 1995). See Box 1.1 and 1.2.

Box 1.1 Common effects of renewable energy

- land use/use of space
- planning/compatibilities
- noise
- visual
- safety
- wildlife

Box 1.2 Source-specific effects of renewable energy

- geomorphic effects (hydro, tidal, wave)
- gaseous emissions (biomass, geothermal)
- liquid emissions (biomass, geothermal)
- soil effects (biomass)
- drainage effects (hydro, tidal)
- eutrophication (hydro)
- subsidence, minor seismic effects (geothermal)

The environmental effects from using renewable energy that are common to all sources are shown in Box 1.1, and include themes such as land use, planning compatibilities and issues concerning minimum standards – for example, of human habitation. Effects which are specific to one or a few particular sources, shown in Box 1.2, can be distinguished, such as soil effects from biomass harvesting or effects on drainage from hydro electric power (HEP) schemes or tidal energy.

Clearly it is possible to identify some common processes and themes despite the very wide variation in the sources themselves and their conversion technologies.

However, there are difficulties in addressing the subject of environmental impacts from such diverse sources as renewable energy flows. While energy is central to the maintenance of natural environments, it cannot necessarily be said that all environmental impacts are energy-related. This would be to use the term 'energy' in a very wide sense, for example chemical pollutants are chemically reactive, and in the strictest sense this could be narrowed down to the ionic charge of the molecules, which could be termed an energy imbalance. An example could be nitric oxide acid atmospheric pollutants. However, the amount of energy per molecule is tiny, and overall the actual energy flow from combustion that goes into creating nitric oxides would also be small. It could still be argued that the reactive power per molecule is relatively high – that is, the density. Notions of energy can thus apply in a certain manner to pollutants which are unwanted by-products.

Central to the natural environment, as a life support system, are the systems and processes involved in cycling the major physical resources such as the well-known hydrological cycle, the carbon cycle, nitrogen cycles and so on. It is energy that drives these systems, the energy of the natural flows listed above. Clearly, intervening in these cycles to any great extent could have consequences.

Environmental impacts also concern the maintenance of stable conditions such as the pH of water, atmospheric pressure and composition, as well as temperatures within the bounds of approximately –50 to +50 °C. Such environmental conditions may not all be primarily directly related to energy flows. Other impacts, such as the reduction of biodiversity from biomass plantations, are not essentially energy-related, being biological and ecological in nature.

1.3 Need for a general theory

Although there has been considerable and growing work studying the environmental effects of using renewable energy sources, the impacts have not been well defined, even though they have been listed comprehensively. Few general principles have been suggested or established. Renewable energy sources offer the prospect of reduced environmental impacts compared with conventional sources, in particular fossil and nuclear sources. However, renewable sources do have impacts, and while these may be relatively insignificant at low levels of deployment, they may well have a cumulative effect. In terms of environmental impact, 'threshold' levels of application may apply beyond which certain sources should not be used, for example at greater densities, if excessive impact is to be avoided. It will be important to establish general principles as the deployment of renewable energy sources increases. In particular, there will be a need to predict and plan for large-scale development, as capacity increases to a point at which the majority of the resource is exploited. It will be important, then, to establish limits to total exploitation based on the effects and impacts as well as the wider issues of economics and technical feasibility.

As capacity increases, and more of the potential is exploited, sensitivities may emerge that were hitherto imperceptible. An example may be the issue of windfarm siting and landscape protection, which has been commented on by McKenzie-Hedger and others. For example McKenzie Hedger states 'how the lack of a coherent policy and assessment framework can lead to difficulties for a new industry, as well as planning departments' (McKenzie Hedger 1995).

Comparisons *between* different renewable sources are required since most locations have the possibility for more than one energy source. Some renewable energy sources are very site-specific, for example hydro electric power, while others are not, e.g. solar photovoltaic. The renewable energy sources themselves are highly diverse and their technologies yet more so, hence comparison on a 'like for like' basis can be problematic. For policy makers, renewable energy sources may be considered together as a group, offering the means of delivering environmental objectives, though in reality the character of the technologies and the industries supporting them may be very different. Many commentators have long pointed to the difficulties posed in trying to compare such diverse sources and technologies – for example, Holdren *et al.* talk about an 'apples and pears' problem (Holdren *et al.* 1980).

Many commentators have drawn attention to the *diffuseness* of renewable energy flows, particularly in relation to concentrated fossil fuel stores. But just how diffuse are the renewable energy sources? Other commentators, such as Walker, have pointed to the importance of land use for renewable energy in 'Energy, Land Use and Renewables' (Walker 1995; MacKay 2008). Both the use of land per kilowatt of output and the issue of compatibility of different land uses are important here.

The first topic, land use per kilowatt of output, suggests that consideration of how intense or diffuse a renewable energy flow – or indeed a source – is will be significant. This leads to the notion of power flux density, mentioned above. This parameter might then be adopted as an initial common denominator in the attempt to characterise and compare the different renewable energy sources. Although the issue of amount of land used per unit of energy could be thought of as a fundamental parameter (MacKay 2008), this is not to suggest that all the environmental impacts can be explained in terms of 'power flux density'. Nevertheless, it may transpire that this parameter, power flux density, has a more pervasive effect on the nature of the source and the technology used to harness it, and therefore on its impacts. This subject is pursued in Chapter 3, where the development of the theory is explained.

1.3.1 Hypothesis

The starting point for the hypothesis, i.e. consideration of the diffuse nature of renewable energy sources, has been outlined above. The hypothesis can be phrased as a research question:

> Is the environmental impact from renewable energy sources related to the power flux density?

The author recognises that this is a very broad question and it is beyond the limited ability of a single writer to answer this question. Of necessity, therefore, what follows is a preliminary analysis. The hypothesis is that impacts arising from renewable energy are related to the power flux density and changes in power flux density, and this is the subject of this study. However, if there is a relationship it is unlikely, except perhaps in some respects, to be entirely straightforward. What follows below is an attempt at analysing the effects and impacts from using renewable energy, which as stated, is a very broad area. Of necessity, this is a multidisciplinary subject straddling the fields of energy, technology and engineering, as well as the biological, geographic and earth and natural sciences and social environmental studies.

In short, the development of the theory as explained below leads to a further three parameters being proposed for a more complete understanding of the processes involved in environmental impact:

1 the proportion of the flow abstracted;
2 the efficiency of energy conversion;
3 the number of energy conversions.

(Clarke 1993)

While this may appear to complicate the assessment model, it should enable further explanation and hopefully prediction of impacts, even for novel renewable energy sources and processes.

1.3.2 Site specificity versus general principles

There is an argument found in the literature that identification of general principles for the environmental impact of renewable energy sources is not feasible due to the site-specific nature of each development, as well as the variety of sources and technologies employed. Indeed, there can be wide variation in the impacts of the same type of renewable energy developments – for example, for hydro electric power plants. In reality, though, these impacts will be explicable, given the effort and will. Every individual site will have a unique mix of environmental characteristics but these are unlikely not to have been encountered before. Rather more likely, it is often the paucity of developments for a particular source, for example of full-scale tidal barrage plants, that has prevented proper experience being gained. Even so, the range in variation between different projects needs to be investigated. The more unique a site is, the higher its perceived environmental value is likely to be. Sensitivity levels to conditions may then be more critical. On occasion this might be due to higher power flux densities – that is, more intense flows of energy – and such sites may then be those most attractive to renewable energy developers. If renewable energy developments are to be considered more than one-off projects – that is, deployed in quantity – it will be necessary to devise common parameters for assessing impact based on sound scientific principles and processes. This will mean recognising site-specific variations in the context of general practice rather than variations being thought of as a barrier to characterisation.

In the author's experience there may even be some resistance to the notion of general principles and common measures for assessing impacts from renewables, from those who are currently practitioners in the field of environmental assessment, who may feel threatened by new techniques. It may be argued that assessment of impact is subjective, or in some way a 'black art'. Other groups who might have tended towards scepticism about the viability of general environmental impact principles may have been some engineering and developer groups who, in the past, might traditionally have regarded environmental issues as barriers and environmentalists as 'the opposition'. Such attitudes have been on the decline in recent years with the growth of a body of well-founded knowledge, though in the past there has been a tendency among some to deny scientific or academic credibility to the environmental sciences.

As stated, some arguments against the identification of general principles centre on the idea that environmental assessment is a subjective issue, and not only is every site unique but so is every person in their predilections. Such arguments are often based on an interpretation of the 'environment' mainly in social or aesthetic terms and thus beyond the more objective measures.

The search for common patterns is vital to a learning process which has as its aim a partial reconciliation of the apparently contradictory objectives of (1) energy abstraction from natural energy flows and (2) nature conservation.

1.4 Conclusion

It can be concluded that as renewable energy source development and resource exploitation proceeds, the ultimate constraints on renewable energy sources will tend to be resource size and environmental ones. The environment can be defined as comprising the natural living environment, the natural inanimate environment and the human or cultural environment. Environmental impacts and effects can be defined as changes. Natural energy flows maintain the natural environment, and a set of functions in nature can be identified for each natural energy source. Since natural energy flows power the natural environment, there will be a need to determine the prime environmental limiting factors. This study concentrates on physical rather than social environmental effects of renewable energy. Some common effects and some unique to certain sources can be identified. There is a very wide variety and diversity of the different renewable energy sources, making understanding of the principles of environmental impacts complex and difficult. Nevertheless, as dependency on renewable energy sources increases, it becomes increasingly important to understand the fundamental principles of environmental impact. It is considered by this author that currently impacts are not well characterised and understood. Since different renewable energy sources may be available in one area, and may be in competition with each other for certain resources, it is important to have a reliable means of comparison. There is thus a need for a general theory of the principles of impact from renewable energy sources. As yet some have argued against the validity of general principles on the basis that site specificity and sensitivities are too varied to make this practical. However, common parameters will need to be devised if renewable energy developments are to be made in quantity.

Bibliography

Blowers, A. and Smith, S. (2003) 'Introducing Environmental: The Environment of an Estuary', in Hinchcliffe, S.J., Blowers, A.T. and Freeland, J.R. (eds) *Understanding Environmental Issues*, Wiley and Open University, Milton Keynes.

Boulding, K.E. (1956) 'General Systems Theory: The Skeleton of Science', *Management Science*, Vol. 2 No. 3.

Boyle, G. (ed.) (1996) *Renewable Energy Power for a Sustainable Future*, Oxford University Press with the Open University, Milton Keynes.

Catlow, J. and Thirlwall, C. (1976) 'Environmental Impact Analysis', Department of the Environment Research Report II, HMSO.

Clarke, A. (1993) 'Comparing the Impacts of Renewables: A Preliminary Analysis', TPG Occasional Paper Technology Policy Group 23, Open University.

Clarke, A. (1995) 'Environmental Impacts of Renewable Energy: A Literature Review', Thesis for a Bachelor of Philosophy Degree, Technology Policy Group, Open University.

El-Moattassem, M. (1994) *Field Studies and Analysis of the High Aswan Dam Reservoir*, Reed Business Publishing, London.

ExternE (1995) *Externalities of Energy*, European Commission, Luxembourg.

Holdren, J.P., Morris, G. and Mintzer, I. (1980) 'Environmental Aspects of Renewable Energy Sources', *Annual Review of Energy*, Vol. 5: 241–9.

MacKay, D. (2008) 'Sustainable Energy without the Hot Air', Cambridge UIT, online resources: www.inference.phy.cam.ac.uk/sustainable/book/tex/cft.pdf, last accessed 20 July 2009.

McKenzie-Hedger, M. (1995) 'Wind Power: Challenges to Planning Policy in the UK', *Land Use Policy*, Vol. 12 No. 1: 17–28.

Mortimer, N. (1991) 'Energy Analysis of Renewable Energy Sources', *Energy Policy*, Vol. 19 No. 4: 374–85.

Petts, G.E. (1984) *Ecology of Impounded Rivers: Perspectives for Ecological Management*, John Wiley & Sons, Chichester.

Twidell, J. and Weir, A. (1986) *Renewable Energy Resources*, E&FN Spon, London.

von Bertalanffy, L. (1968) *General System Theory: Foundations, Developments, Applications*, Braziller, New York.

Walker, G. (1995) 'Energy, Land Use and Renewables: A Changing Agenda', *Land Use Planning*, Vol. 12 No. 1: 3–6.

2 Literature review

2.1 Introduction

There is a considerable quantity of literature on the subject of environmental impacts from renewable energy, dating from the 1970s to the present time.

The author has reviewed a selection of the literature on environmental impacts from renewable energy sources (Clarke 1995), as well as works on individual sources such as wind (Clarke 1988) and biomass (A. Clarke 2000). It is not proposed here to review again all the individual impacts of each renewable energy source, but to concentrate upon key developments in the approach to assessing impacts from renewable energy.

Since the 1990s much has changed as renewable energy utilisation has grown, and emerging technologies, e.g. marine current or wave, have been brought into or near to commercial operation. Environmental awareness has increased among the public, the energy industry and governments. Scientific knowledge of the effects of energy use has increased greatly, while evidence for anthropogenic climate change has consolidated. The privatisation of electricity and energy markets in the UK has resulted in some renewable electricity being sold competitively on the basis of its environmental credentials, as 'green power'.

The rise in energy prices over the last 10 years with increasing volatility, led by the rise in oil prices from ~$10 to a peak of almost $150 per barrel by July 2008, falling thereafter to ~$40 as a result of recession, and then rising to over $100 for about five years before falling once more, has renewed the urgency for government policies to plan for a post-oil and post-fossil fuel age.

Together with increasing evidence for the potential severity of climate change effects, this has led to policies for strong renewable energy supply growth; in turn this has caused further examination of the size of total resources available and their constraints.

2.1.1 Recent developments

Since the 1990s, much experience has been gained of new renewable sources on a larger scale than previously – for example in wind energy, and in many countries biomass energy. In the UK, installed wind energy capacity is over 11 GW, generating on average about 8 per cent of UK electricity at the time of writing, while Denmark was using it for about 28 per cent of electricity generation (EWEA 2013) and Germany had 22.247 GW, contributing over 11 per cent of the electricity supply. Wind capacity installation has been expanding worldwide at a rate of 31 per cent per annum, and surpassed 100 GW in 2008 (EWEA 2008) and was over 336 GW by June 2014, generating about 4 per cent of world electricity. Despite the expansion of wind energy in the UK, there has been effective opposition on environmental grounds, which has slowed down development.

Another significant change since 1995 is that the emphasis of environmental policies has been put more firmly on carbon dioxide reduction and greenhouse gas (GHG) control. As such, CO_2 and GHGs serve as a proxy for other impacts.

This emphasis in energy policy on replacing fossil fuels, in order to control atmospheric carbon emissions, has resulted from the increasing confirmation by the Intergovernmental Panel on Climate Change (IPCC) of the warming effects of the continuing growth of CO_2 and other GHGs in the atmosphere (IPCC 2007). Although considerable uncertainty remains about the rate of climate changes to be expected, and models are not yet considered to be capable of accurate forward forecasts, there is a growing consensus that warming due to accumulation of GHGs in the atmosphere will cause climate changes and melting of ice, leading to further changes, e.g. in sea level.

As a reaction to this growing consensus on climate change, government policies have been implemented, with regulation addressing global warming, including renewable energy policies, in addition to more general environmental regulation, e.g. on air pollution. Accompanying this has been the development of assessment and accounting techniques for carbon, and its environmental cost, such as those pioneered by the ExternE project (ExternE 1995a), for CO_2 and GHG pollution by energy technologies, directed towards government policy areas (WEC 2004). The unit of measurement used is grams of carbon dioxide equivalent per kilowatt-hour of electricity generated, or gCO_2eq/kWh.

The growing maturity of some of the renewable energy sources, such as onshore and offshore wind, and some biomass options especially in Continental Europe, has led to renewable energy sources being seen as a serious electricity and energy supply option by governments in, for example, the EU and the UK. As an example of how seriously renewable energy sources are now taken, the government's 2009 'The UK Low Carbon Transition Plan' (DECC 2009a) allocated them a major supply role. The emphasis, then, has been on achieving government targets for CO_2 and GHG reductions, and raising the capacity of low carbon and renewable energy generation. The Climate Change Act of 2008 requires a reduction in emissions of 80 per cent by 2050 compared with 1990 levels. A reduction to 50 gCO_2eq/kWh for electricity generation by 2030 has been recommended by the Climate Change Committee.

At the scale of individual renewable energy schemes, there has been considerable further development of local impact assessment techniques for individual developments – for example, software for individual windfarm development impact assessment and site evaluation (Alan Harris of Resoft, personal communication 1999; Garrard Hassan, personal communication 2004; *Windpower Monthly* 1998).

While much has changed and there has been considerable progress on various fronts, some of the original uncertainties and problematic aspects of renewable energy sources still remain: practical resource size, consequences for land use, and the degree and extent of impact from what are very diverse and relatively novel technologies, many of which have not been exploited on a larger scale.

2.2 Key issues in the literature

Holdren *et al.* (1980) were some of the first to discuss the problem of comparing the effects of different energy alternatives for renewable sources in a much-quoted paper, 'Environmental Aspects of Renewable Energy Sources'. They state in an introduction that begins with the subject of 'externalities' that 'the environment is central to the energy problem, not peripheral'. The paper considers what information is required for comprehensive comparative assessment of energy alternatives, summarising the technological characteristics of the most promising renewable energy technologies, emphasising how they cause environmental problems, comparing and offering suggestions for the implications of energy choices. The elements of a systematic environmental assessment are described as steps in tracing the complexity of causal linkages: origin of environmental effects, insults to immediate environment, pathways by which insults lead to stresses, stresses resulting in altered environmental conditions, and damages which are the response of components to stresses.

A complete analysis would need to contain, the authors state, information on: the distribution of damage in space and time, ease of control of the damage, degree of irreversibility, how the damage scales and degree of uncertainty. The authors acknowledge that such complete information is not available even for well-studied technologies. The 'most intractable problems in environmental science' posed are the processes linking insults to pathways and stresses and their relation to damages. The key point that renewable energy flows are diffuse is made, and that this results in greater land use and materials use. In addition, the very important point that renewable energy flows power the environment and biosphere and 'large enough interventions in these natural energy flows and stocks can have adverse effects on essential environmental services'.

Of all the alternative sources compared – passive and active solar, solar thermal electric, photovoltaics, terrestrial and orbiting solar satellite, wind, hydro and ocean thermal energy conversion, hydro electric power is considered the worst option due to its ecosystem damage per unit of energy. The increasingly scarce resource of 'free flowing rivers' is pointed out.

Some basic measures of relative severity of environmental effects are suggested: land use, water use, non-fuel materials, occupational accidents and diseases, public risks from accidents, effects of routine emissions on public health, effects on climate and ecosystem effects. Although the effects of dispersed renewables are considered to be only local, if large enough deployment of centralised renewable sources occurred, the effect of redistribution of energy flows would become appreciable, the authors state. Ecosystems, while being valued as part of 'reverence for nature', also provide goods and services, which contribute to human wellbeing, e.g. nutrient recycling, soil formation, water storage and flow regulation, as well as maintenance of a genetic library (biodiversity).

However, although there is a large body of observational data on ecosystems, the authors consider that few general principles are known. This is explained by the 'tremendous complexity and stochasticity of such systems which makes prediction of human induced stresses extremely difficult'. Therefore they state, 'comprehensive comparison of different energy technologies is at a necessarily very primitive stage, being qualitative rather than quantitative'.

Another influential work, the OECD Compass Project Report 'Environmental Impacts of Renewable Energy' (OECD 1989), considered that 'renewable energy technologies can in general be considered as more environmentally favourable than most other sources', but they have different impacts. It states that because 'renewables tap more dilute energy flows which are generally of a more physical rather than chemical nature, the impacts of renewables tend to be more physical rather than chemical'. The subject of the larger land use and materials inputs of renewables is discussed. However, important positive aspects of renewable energy use are that 'renewables generally imply more localised, shorter term environmental effects'. While the report considers a range of renewable energy technologies, the authors state that 'at present, methodological tools to enable a truly rigorous analysis . . . are not available'. This, they believe, is due to the inability to identify system boundaries, which are particularly problematic for renewables. Although it is possible to mitigate the impacts of renewables,

> unfortunately past experience suggests that the full impacts are only recognised after adverse effects have reached significant levels, because insufficient attention has been given to predicting harmful effects in advance and remedies incorporated into the maturing technology as an integrated system.

An influential group of commercial and academic specialists, the Watt Committee, published a comprehensive report, 'Renewable Energy Sources' (Watt Committee 1990), which as well as reporting on the status of the technology and prospects for exploitation in the UK, considered environmental impacts of energy production, both as a stimulus and as a barrier. They state that renewables are popularly thought to be an environmental improvement, but can introduce impacts of their own, such as visual intrusion, noise and radio interference from wind turbines, toxic emissions from biofuels, interference to fish and water flow from hydro, conflict of land use, and other disturbance to the natural habitat. The authors significantly note that 'no satisfactory method of comparative assessment in this area exists'. They acknowledge that environmental considerations have become more important in energy policy due to the effect of global warming and climate change through CO_2 and other GHG emissions. As a result 'renewable energy technologies thus find themselves in a most encouraging position as far as research development and demonstration policy in IEA member countries is concerned'.

Once again, only qualitative comparisons are made, devoted to impact summaries between renewable and conventional sources, showing the relative magnitude of effects as small, medium or large. These emphasise the advantages of renewable sources over conventional ones, as the only impact shown which is greater for renewables than for fossil or nuclear is land use, due unsurprisingly to its lower energy density.

A somewhat more recent contribution from a conference held by the UK Solar Energy Society 'The Environmental Impacts of Energy Technology' (Hill and Twidell 1994), included presentations of papers on photovoltaics and the environment, as well as the role of energy efficiency in reducing impacts and environmental cost assessment of conventional electricity generation. Hill outlined many of the themes referred to below and cites some of the papers above.

Twidell, in a paper 'The environmental impacts of wind and water power' (Hill and Twidell 1994), proposed five categories of environmental impact: physical, chemical, biological, ecological and aesthetic. Twidell referred to energy flux density at first capture as a factor with strong influence on physical and aesthetic impacts, though he did not develop this very much. He noted, however, that hydro power has about 100 times the energy flux density of wind and solar, and that another factor is the efficiency of conversion. Twidell adopted a similar approach in some respects to that of the present author in an earlier paper (Clarke 1993).

Since the middle of the 1990s, efforts have been made to develop comparative assessment techniques that rest upon quantitative data rather than the qualitative descriptions prevalent until then. This process has continued with, for example, government agencies, the Carbon Trust, the Sustainable Development Commission, the Tyndall Centre and the UK Energy Research Centre, as well as university researchers, industry consultants and developers, developing assessment techniques.

As these technologies begin to be deployed and enter the planning system, the focus tends to be on the potential social and environmental impacts and implementation problems of specific energy options, such as on-land windfarms and more recently offshore wind, wave and tidal energy systems. An example is the Strategic Environmental Assessment of offshore wind technology (DECC 2009b).

Some of these studies may feed into wider comparative impact assessments, but as yet there is no reliable methodology for making comparisons, although some of the techniques used in the studies may provide a starting point.

2.3 Assessment techniques

2.3.1 Cost–benefit analysis

The difficulty of assessing impacts of very diverse renewable energy sources has been addressed by a variety of assessment techniques. Underpinning many of these is cost–benefit analysis. Cost–benefit analysis attempts to measure all of the benefits and costs of a project or activity, in financial terms, and to allow the project to proceed if the sum of the benefit exceeds the sum of the costs by a sufficient margin (Cross 1982). It aims to provide monetary values for social and environmental effects and damages which are not reflected in the normal marketplace and therefore do not have an allocated cost in monetary terms. Such costs are effectively ‘external’ to the normal market valuation process which allocates monetary values and are termed ‘externalities’. The technique relies on listing costs and benefits, allocating them a monetary value and then assessing the ratio. Thus more can be taken into account than in a purely economic appraisal since the value of social and environmental factors can be put into the balance. For example, pollution effects such as the damage caused by acid rain can be valued and a cost allocated. The technique can be extended to renewable energy sources such as wind energy, where, for instance, loss of visual amenity can be costed, as has been done for Cornwall, where a 1.9 mECU/kWh cost was cited as an upper limit (ExternE 1995b: 80).

Costing and valuation assessment techniques used previously have often relied on cost–benefit analysis as a way of reconciling the great variations in the different renewable energy sources.

As with systems theory, a weakness of cost–benefit assessments lies in the determination of where the boundaries are drawn. What is included and what is excluded can alter the result significantly. Additionally, the results will only be as good as the data and models adopted. There can be great difficulties in allocating monetary values to certain impacts such as aesthetic factors or biodiversity which at present are either very uncertain or even indeterminate or unknowable.

2.3.2 Life cycle analysis

Another technique that has matured and come into widespread use is life cycle analysis (LCA) (also termed life cycle assessment and life cycle inventory), which is a method of evaluating processes, products and activities through their life cycle '*from cradle to grave*' (EEA 1998). LCA traces the whole of the cycle of materials, energy and processes involved in products and activities, from the raw materials in nature through manufacture and use to final disposal (Royal Society of Chemistry 2005).

The first stage involves quantifying the raw material acquisition, the natural resources and energy inputs, and the outputs, consisting of air emissions, solid waste and wastewater. The next stage analyses material manufacture; a third stage considers product manufacture; and in the fourth stage product use is analysed. The last stage considers product disposal. For each stage, inputs from the previous stage are included and energy inputs and outputs of air emissions, solid waste and wastewater. These stages are common to most LCA exercises, but can be extended or modified for particular cases (EEA 1998).

After constructing a flow chart, the inputs and outputs of each stage need to be identified and listed for the inventory. This will record the energy and materials inputs and outputs for the particular product, process or activity. Since not all the inputs and outputs in LCA can be identified, it is necessary to select the main ones. Standard units need to be used for the mass and energy tally. A common framework is required for comparison and data are kept to a minimum.

LCA is an accounting tool that uses standard data to indicate pollution and impact associated with emissions and resource consumption, which may not apply in specific cases. Although LCA appears to convey an impression of precision, in fact it can only convey an indication of the likely impacts (Jensen 1997). The scope of the exercise needs to be carefully determined, with any gaps and omissions needing to be taken into account.

LCA and assessment are used to compare different products, processes or activities, to identify the main inputs and outputs and their impacts, including pollution, and to determine which stages in the life cycle cause the main impacts. As such, LCA provides a type of environmental sensitivity analysis. A choice of options can then be pursued to reduce overall impact.

LCA tends to be most successful when used for products or elements of processes that can be easily delineated. For example, when applied to the individual energy converter devices such as wind turbines or solar photovoltaic modules, it can help highlight which stages of their life cause the most impact. Where systems are more complex or open-ended, involving wider cycles, the narrow focus can, however, be misleading. LCA can nevertheless provide a useful analytical tool when used properly. The International Standards Organization (ISO) has developed a standard for LCA: ISO 14042, 14040, and 14044 Life Cycle Impact Assessment standard (ISO 2006).

An example of a comparative LCA for energy systems is the study by the World Energy Council (WEC 2004), which seeks to quantify the key impacts for all energy sources, including renewables, though mainly in terms of GHG emissions.

LCA has been used extensively in assessing impacts from energy, and in 'internalising the externalities', i.e. bringing into the cost realm impacts which have formerly been unaccounted for. For example, land use impacts, which lie at the root of the raw material stage, can be incorporated in a framework which addresses such issues as biodiversity and soil quality (Mila i Canals *et al.* 2006). The methodology of LCA is used by the ExternE project, reviewed below.

Criticisms of LCA methodology which may not take into account changing energy mixes are addressed by Hertwich *et al.* (2015). They used LCA in a study to investigate the environmental benefits of low-carbon technologies, described in a paper titled 'Integrated life-cycle assessment of electricity-supply scenarios confirms global environmental benefit of low-carbon technologies.' The authors use what they claim to be 'the first global integrated life-cycle assessment (LCA) of long term, wide-scale implementation of electricity generation from renewable sources, (i.e. photovoltaic and solar thermal, wind and hydropower) and of carbon dioxide capture and storage for fossil fuel generation', 'to assess the tradeoffs of increased up front emissions and reduced operational emissions' of renewable sources. Their methodology takes into account 'newly installed capacity year by year for each region, thus accounting for changes in the energy mix used to manufacture future power plants'. They conclude that low-carbon sources allow a doubling of electricity supply while stabilising or even reducing pollution, even though 'material requirements can be higher than for conventional fossil generation: 11–40 times higher for copper for PV systems, and 6–14 times higher for wind power plants'. The authors analyse the IEA Blue Map scenario (IEA 2010) and compare it with the IEA's Baseline scenario. Emissions causing particle matter exposure, freshwater eco-toxicity and eutrophication and climate change for the climate change mitigation (Blue Map) and the business-as-usual (Baseline) scenarios up to 2050 are compared.

In addition to the technique of allocating a monetary value to all of the effects, as in cost–benefit analysis, another approach to the problem of comparing entities that are incommensurate is known as multi-criteria analysis (MCA).

Bucholz *et al.* state in a paper, 'Multi Criteria Analysis for bioenergy systems assessments' (Bucholz *et al.* 2009), that 'MCA can be defined as "formal approaches which seek to take explicit account of multiple criteria in helping individuals and groups explore decisions that matter"' (Belton and Stewart 2002; cited in Bucholz *et al.* 2009). This approach stands in contrast to the 'single goal optimisation approach', with 'unifying units', (e.g. cost–benefit analysis), as the authors state.

A variety of methods have been devised and MCA can be classified as a multi-objective decision making (MODM) approach, where an indefinite number of scenarios or sets of principles are involved. The degree of fit of the principles with the scenarios is given a numerical value. Each principle can be weighted to reflect how it is valued by the community.

Bucholz *et al.* assess four different MCA tools on the basis of their ability to help design and plan for sustainability in a bioenergy project in Uganda (Bucholz *et al.* 2009). A 10 kW generator, wood gasifier plus plantation, supply chain and mini-grid scenario was tested against existing individual petrol- and diesel-powered generators. Land area used, at 3–14 ha, depending on productivity as well as gasifier efficiency, still resulted in a competitive price against kerosene, petrol and diesel. The use of a 'radar graph' showed, perhaps unsurprisingly, that the fossil fuel scenario resulted in decreased competition for fertile land (Bucholz *et al.* 2009).

The advantages of using MCA, according to the authors, are that it can structure the problem, it can assist in identifying the most uncertain and least robust parts and, lastly, it can bring stakeholders into the decision-making process. However, the authors state that considerable variation in results was obtained from the four different MCA tools, though social criteria were identified by all as being decisive.

While MCA can offer one solution to the problem of comparing 'apples and pears' in that social criteria can be weighed together with economic criteria, it can still rely on generalised parameters, such as in the case of 'reduced pollution'. Average values are employed and their range can be problematic. It is perhaps best suited for political decision making, involving such social factors as the amount of employment produced (Bucholz *et al.* 2009).

EROEI

EROEI is an energy accounting method and stands for 'Energy Return on Energy Invested'. The method has been used to determine the productivity of energy conversion devices in energy terms. In order to determine the EROEI, the energy return is divided by the energy invested so that:

$$\text{EROEI} = \frac{\text{Energy return}}{\text{Energy invested}}$$

(Humphreys 2013)

The EROEI needs to be greater than 1 for the technology to be practical, and this metric has environmental implications too, as EROEI is a measure of efficiency. In measuring the EROEI of processes, only human-applied energy inputs are counted, i.e. the energy input into biofuels from solar radiation is not included, nor is the waste energy. The work of Mortimer (1991) pioneered energy accounting for a number of energy technologies, though the term EROEI is a later convention. EROEI can be applied to all energy sources and it has been argued that industrialisation and development of higher living standards depended on energy sources with greater EROEI values. The fate of whole civilisations has been linked to declining EROEI, e.g. with regard to the Roman Empire (Homer-Dixon 2007). An example of the decline of EROEI is the exploitation of early oil reserves which had EROEI values of nearly 30 (Fridley 2010), but as resource depletion occurs the EROEI values of new oil finds have declined. This is particularly the case with oil shale and unconventional oil sources. Murphy and Hall (2010) describe an EROEI ratio curve for energy sources, with biomass sources barely above 1, ethanol from sugar cane in single figures equal to shale oil, solar PV at about 5–6 EROEI, nuclear at about 10, wind at about 18, oil production at 20, while oil and gas in 1970 are shown at about 30, world oil production at about 35 and coal at 80, with hydro electric power (HEP) at an EROEI of 100. However, other sources give different estimates of EROEI, and there is little agreement on the methodology. For example, there is no agreement on where the boundaries should be drawn in terms of energy invested 'upstream' in the supply chain, at the factory or all the components and raw materials and processes of the factory and wider infrastructure. In this respect the concept is similar to LCA and may be used as part of LCA analysis.

EROEI has been called 'a quintessential but possibly inadequate metric in a solar powered world' (Pickard 2014), as the returns from, e.g. solar PV, accrue over time, but the value may not be fully accounted for, as the form of the energy generated is not taken into account. High-quality energy forms such as electricity are more valuable than low-quality forms such as heat; chemical energy forms such as fuels are valuable if their energy density is high, and, e.g. in liquid form. Solid forms such as coal are less valuable, since conversion for other forms is required – usually at low efficiency – and they are bulky and polluting (Boyle 1996: ch. 1). Low-carbon energy sources' value is not taken into account, e.g. electricity is treated in the same terms as the primary chemical energy of coal.

Moreover, EROEI may have more to say about human endeavour, i.e. economics, rather than a fundamental measure of potential efficiency or intractable barriers, since in the example of renewable sources EROEI, such as solar or biomass, the natural energy flows are not accounted for. The technique has certainly been useful as a measure of fossil fuel reserves, particularly oil, where the EROEI has risen as drilling and exploration has extended to the ocean floor and the Arctic. However, EROEI may only provide a snapshot of the state of development of an energy technology – admittedly useful, but early studies of the EROEI of solar PV merely point to a problem of immature technologies, in that devices and processes require more development to attain greater productivity and higher efficiencies. Richards and Watt (2007) used an Energy Yield Ratio rather than Energy Payback Time to estimate higher values of EROEI at 4.8–13.9, which they state 'is many times the energy used to fabricate the system'. Solar PV is still an immature technology, and its conversion efficiency is rising, and production volumes are rising, so the EROEI value should be rising too. EROEI is undoubtedly a useful metric, but one which has limitations too.

UK government agencies have been involved in assessment of environmental and social impact studies such as the large scale pan-European ExternE study (see below), which aims to cost the impacts of various energy sources so that comparisons can be made and assessed on a cost–benefit basis.

2.4 Externalities of energy project: ExternE

A large international collaborative study by the European Commission and, until 1995, the US Department of Energy commenced in 1991 (ExternE 1995a), aiming to be the first systematic approach to the evaluation of external costs of a wide range of different fuel cycles. It aimed to provide a transparent basis on which different impacts, technologies and locations may be compared.

The external costs of energy are those not accounted for by the market, i.e. environmental and social costs. A multi-disciplinary approach was taken, including health, ecology, materials science, energy, economics and atmospheric modelling. The two types of information sought were quantification of impacts and then the valuation of these impacts under a common measuring system.

The accounting framework comprises the Fuel Cycle stage, resulting in an Activity; the Activity results in a Burden; the Burden results in an Impact; and finally the Impact itself, which leads to Valuation.

On the question of where system boundaries are drawn, this may be 'upstream', e.g. impacts of steel-making are included if they are significant, but upstream impacts are not taken to a logical infinitesimal extreme. The authors state that 'Ideally impacts should be assessed over full life time and full geographical range.' In this manner the ExternE is a version of LCA.

The methodology used is the 'impact pathway methodology'. Impact Pathway Methodology involves defining the impact pathways, identifying the models of impact, the pathway stage, and the reference environment characteristics (ExternE 1995a). The ExternE summary describes this method as 'bottom up' as opposed to most published studies, which are 'top down', i.e. using average data, not related to the location or time. Site-specific data are used and applied in impact pathway methodology.

Krewitt, reviewing the ExternE Project in a paper (Krewitt 2002), concludes that the answers provided by the ExternE study do indeed match the questions, but are only partial answers, reflecting the complexity of the environment, though he acknowledges that this is a big achievement in itself. He recognises the reputation and achievements of ExternE in providing a 'benchmark' standard for studies of externalities.

His criticisms concern the uncertainties and limitations of external costing, e.g. the robustness of valuations, such as 'Value of a Statistical Life' (VSL), the coverage of key impacts, and external cost estimates in a policy context. VSL concepts needed to be modified for the case of shortening of life expectancy, as opposed to risk of mortality. The coverage of key impacts such as climate change from global warming, resulting in minimum values of €0.1 to a maximum value of €16.4, given the uncertainties involved, are considered 'at least badly misleading'. Krewitt considers that ExternE also fails to recognise the reversibility of an impact adequately. On the issue of 'beyond design accidents' and nuclear waste impacts over huge time spans being positively discounted, this '"awards" the irreversibility of the effect, by bringing the monetary value of the long term effects practically down to zero.' This is a fair point, though there may be confusion about the type of wastes involved and the wide variation in half lives. Intermediate-level nuclear wastes are indeed much easier to deal with after long storage periods.

Krewitt points out that on renewable energy sources, the variability of the impacts from the material intensity of renewable energy technologies reflect excessively the energy mix of the country of manufacture (i.e., high or low fossil fuel/carbon, e.g. Germany versus Switzerland), rather than adopting an average representative value. Nor, does he believe, site specificity is adequately acknowledged for renewables with regard to the limited availability of suitably low-impact sites.

Concerning the policy influence of external cost estimates, Krewitt's view is that despite ExternE failing to calculate sufficiently robust, comprehensive and precise external cost data, external cost estimates were still incorporated into European environmental policy. The magnitude of electricity price 'adders' that are required to internalise external effects, for global warming, has been limited to a maximum of €0.05 per kWh. Krewitt considers that this should be a lower boundary, not a best estimate, considering the limitations of estimations and the uncertainty and magnitude of the issue. However, he points out that the 'relative ranking of energy technologies was much more robust, indicating lower external costs for electricity generation from hydro, wind, and in many cases, from PV compared to fossil fuels'.

2.4.1 Conclusions regarding ExternE

The ExternE study is based on an LCA approach to environmental impacts, which assessed categories of damage, and costs of impacts and effects. The impacts and effects were obtained from surveys of literature in the field. While this approach is reasonable, it relies on existing models of impact. These have largely concentrated on conventional sources of energy such as fossil fuel thermal energy conversion, i.e. combustion, or on nuclear energy. These sources have been so dominant in electricity generation that approaches to them such as 'Fuel Cycle' categorisation appear to be standard and are then applied to all of the sources. The application becomes increasingly tenuous when applied to renewable energy sources, which apart from biomass, are not fuel-based. This emphasises the need for a different model of impacts. Indeed, the whole life cycle approach, i.e. 'from cradle to grave' implies a definite beginning phase, a use phase and then an after use, or waste-product phase. This is more applicable to fuel-based sources of energy than to renewable sources which represent the harnessing of continuous flows of energy. Life-cycle analysis is more suited to products where clear system boundary lines can be drawn. It is, however, a very useful technique in determining, for example, energy payback accounts of discrete energy converter devices and equipment, and can be a powerful tool for assessing total impacts where they can be readily identified.

2.5 'Green' electricity criteria

The advent of 'green' electricity supply, sold as deriving from solely renewable sources and charging a premium, has led to some examination of the criteria for 'green' generation and even what constitutes 'renewable' (A. Clarke 2000). For instance, the question of whether waste sources are truly renewable and how 'green' they are can be posed, given that waste streams incinerated for energy purposes can include plastics and other fossil fuel-based materials.

For an example of this as applied to hydropower, the EAWAG, Swiss Federal Institute for Aquatic Environmental Science and Technology Oekostrom Greenhydro project has issued international guidelines for green electricity using hydropower (Bratrich *et al.* 2000). EAWAG has developed a certification procedure for green power from hydro sources using hydropower criteria, such as minimum (compensation) flow criteria, concerning the diversion of water, as well as storage. Also included in the criteria is interconnection between watercourses, groundwater, riparian zone and floodplains, adequate water for fish migration, as well as preservation of natural river beds. The criteria extend to hydro peaking, i.e. the use for peak demands, also reservoir management, sediment transport bed load management, and power plant design criteria. Bratrich *et al.* (2000), make the point about 'green electricity' that 'there is still a lack of credible guidelines for the certification of such products', and that such products have tended to concentrate on the new renewables (excluding hydropower). Even where such sources are included, while the global criteria were met, local impacts to the river were frequently not considered. The Swiss situation was cited where some 80 per cent of the technical resource capacity for HEP is already exploited: 'The effects on the ecological integrity of many alpine river systems are dramatic.' The interesting proposal here is to use the surcharge for 'green' electricity to achieve higher standards in HEP generation, through the Ökostrom (eco current) project (Bratrich *et al.* 2000).

The author has posed the question of whether some green energy sources are greener than others, and if so by what criteria, in an article, 'Buyer beware' (Clarke 1998).

As an example of particular land use problems, and the competition with existing uses of farmland, as well as the issue of target-setting in advance of research on the effects of a policy, the biomass and biofuels controversy is included here. Biomass was the largest source of renewable energy in 2004 (Boyle 2004), but also has the lowest power flux density.

2.5.1 Biomass and biofuels controversy

The issue of target setting for renewable energy sources, and for biomass and particularly biofuels policies, had become environmentally controversial by 2008. Although there has been environmental scepticism concerning biomass energy usage ever since large-scale policies were first proposed in the 1970s (Pimentel *et al.* 1983), growth in the use of such sources has been gradual until more recent policies were announced, e.g. the EU Biofuels Directive (EU 2003). Rising targets for 'green' biofuels use were set for road transport, resulting from policies to increase renewable energy capacity overall, and have led to claims that competition for land area is driving up food prices, that forest is being cleared for palm oil plantations, for example, and that the carbon savings from biomass energy are much lower than has been claimed (FoE 2010).

Several 'food versus fuel' reports were produced by different organisations. A leaked World Bank draft report (Mitchell 2008) was said to claim that 'Biofuels have caused 75% of the food price rise' (Chakrabortty 2008). In fact, when it was formally released, the report by World Bank economist Don Mitchell said that biofuels, low-grain inventories, speculative activity and food export bans had together pushed food prices up by 70–75 per cent. So it was not just biofuels – it was also trade patterns. Also, a study by New Energy Finance (Liebreich 2008) found that since 2004 biofuels were responsible for at most 8 per cent of the 168 per cent rise in global grain price and 17 per cent of the 136 per cent rise in edible oil prices.

Although the EU Biofuels Directive of 2003 (EU 2003) had encouraged farmers to set aside land for biofuels, there was not very much production in the UK – approximately 120,000 tonnes per annum – which was small compared to whole-world production.

A paper by Tan *et al.*, 'Palm Oil: Addressing Issues and Towards Sustainable Development', in support of palm oil growers, defends the use of palm oil for biodiesel (Tan *et al.* 2009). This is a riposte to WWF reports (WWF 2003), (WWF 2005) on the impacts of palm oil use for biodiesel production, which they claim include deforestation, biodiversity loss, orang-utan extinction and peatland destruction. Tan *et al.* state that palm oil constitutes only 1 per cent of biodiesel feedstock, but produces >4t per hectare on average compared to 1.5t per hectare in the case of rapeseed oil, and at a lower price per tonne. The authors state that the claimed impacts of palm oil for biodiesel on deforestation are exaggerated, as out of 4,050 million hectares most are already cultivated, i.e. production has only shifted from other crops. Palm oil has better gross assimilation of CO_2 per annum, than rainforest, with a high photosynthetic efficiency of 3.2 per cent claimed. Other advantages are a low fertiliser requirement, relatively good biodiversity, with under-cropping possible. Tan *et al.* also make suggestions for sustainability criteria. The biofuels issue became increasingly urgent, given the targets being set for renewable energy and transport fuels.

2.5.2 Targets

EU targets for obtaining 20 per cent of EU energy by 2020 from renewables were instituted in 2007 (EREC 2007; Europa 2008) and also 10 per cent of transport fuels from biofuels by 2020. In the UK, Renewable Obligation targets of 10 per cent of electricity by 2010 (BERR 2007), and 15 per cent by 2015 were adopted. Government targets for road transport biofuels in the UK were set at 2.5 per cent of the total initially, and then 5 per cent by 2010, subsequently increasing to meet the 10 per cent by 2020 EU Target (EU Biofuels Directive 2003).

A Policy Review undertaken by the government, 'The Gallagher Review of the Indirect Effects of Biofuels Production' (Gallagher 2008), recommended reduced growth of biofuel use, effectively halving the rate of growth, pending use of marginal lands and advanced technologies such as enzyme 'cracking' of cellulose (RFA 2008).

Subsequently, the government decided on a 3.25 per cent target for biofuels in UK transport fuel from April 2009 to March 2010 – a reduction in the 3.75 per cent target announced before the Gallagher Review. Subject to Parliamentary approval of the RTFO (Amendment) Order 2009, subsequent targets were:

- 3.5 per cent for 2010/11;
- 4 per cent for 2011/12;
- 4.5 per cent for 2012/13;
- 5 per cent for 2013/14.

The Gallagher Review of the indirect effects of biofuels production looked at the impacts of biofuels especially in land use change and rising food prices. However, although the review found that increasing demand for biofuels 'contributes to rising prices for some commodities, notably for oil seeds', it could not provide estimates of how much since the data were 'complex and uncertain'.

Some criteria for sustainability were nevertheless proposed so that biofuel production does not compete with food production and that carbon neutrality is achieved (Gallagher 2008). This policy report for government addressed concerns over biofuels policy in contributing to food price increases, deforestation and GHGs. The authors concluded that probably enough land area exists for both biofuel feed and food production. Idle and marginal land should be used. Advanced technologies such as second-generation biofuels using enzymes should be stimulated. Biofuels do contribute to rising food prices for the poor, according to the authors, but a sustainable biofuel industry is possible. Lower targets, such as 5 per cent, not 10 per cent, of road fuel by 2020, and stronger controls are needed, they state, together with global enforcement to prevent deforestation.

According to critical observers such as NGOs like Friends of the Earth (FoE 2008), the emphasis on targets and rapid growth before proper sustainability criteria have been developed has led to impacts. It would appear that sufficiently robust criteria, based on models incorporating quantitative data on, e.g. soil carbon balances, were not yet available, leading to unrealistic expectations of biofuels.

Biomass as fuel for electricity generation

However, there has also been controversy over electricity generation, where biomass has replaced coal as a fuel. More recently a report by the Department of Energy and Climate Change, 'Life Cycle Emissions of Biomass Electricity in 2020' (Stephenson and Mackay 2014), has tried to take more complete impacts into account. This provides estimates of CO_2 and GHG emissions from electricity generated by converted coal power stations such as Drax, using wood pellet fuel imported from North America. This study takes into account carbon stored in the reservoir of standing older timber, rotting trunks and possibly soils, assessed over different time spans of 40 and 100 years. The report concluded that if thinnings, sawdust and waste wood from the timber industry were used, there could be GHG savings, but if whole stem (tree trunks) were used as fuel, taking into account the energy and carbon costs of drying, pelletising and transport, there could be higher emissions per unit of electricity than coal. However, the range of emissions in terms of carbon balances was large, depending on the baseline comparison, i.e. the previous land use, here termed 'counterfactual'. The findings of the report indicate that the rate of harvesting and the ultimate fate of the wood significantly influenced estimated emissions.

2.6 General environmental impacts

While in recent years there has been no shortage of adverse and sometimes partisan commentary on local impacts of specific renewables, e.g. by groups opposed to windfarms, and there is a large literature on the issue of public reactions to renewables (Walker 1995a, 1995b; Elliott 2003; Bell *et al.* 2005; Warren *et al.* 2005; Devine-Wright and Devine-Wright 2006; Ellis *et al.* 2006), there have been relatively few attempts at comprehensive comparative assessment of environmental impacts.

In a somewhat pejorative view of impact of renewable energy systems when applied on a large scale, which nevertheless makes some useful points concerning large land and water use, Abbasi and Abbasi (2000) review the negative effects of renewable energy sources. In a paper titled 'The Likely Adverse Environmental Impacts of Renewable Energy Sources', they state that all renewables have impacts, with solar heating and passive solar having least and wind few. Most types of renewable energy systems are reviewed. However, some of the material referred to appears to be selectively chosen and dated, such as the quoting of low-frequency noise from the MOD-1 wind turbine of 1979. The authors write from a largely Indian perspective, in which hydro electric power is considered particularly bad, together with centralised biomass. The low overall photosynthetic conversion efficiency of biomass at 0.1 per cent is cited. The authors state they are not against renewable energy, but caution against it being seen as a 'panacea'.

The present author has, however, written in this field, and reviews several of his own works here.

2.6.1 Previous work by the author

This author has written a series of previous works, out of which this book has been developed. 'Comparing the Impacts of Renewables' (Clarke 1993, 1994) is a theoretical model and analytical structure for the assessment and comparison of impacts of different renewable energy sources. It suggests a predictive model for how impacts are caused based on power flux density, but also introduces other parameters such as proportion of the flow extracted, efficiency of energy conversion and number of conversions. A version as a paper was published in the *International Journal of Ambient Energy* (Clarke 1994), as well as a report version. Although some preliminary figures are used in graphs and tables, the model is essentially uncalibrated, being a hypothetical structure for the mechanism of impacts.

'Environmental Impacts of Renewable Energy: A Literature Review, Thesis for a Bachelor of Philosophy Degree' (Clarke 1995) aims to be a comprehensive review of the impacts of renewable energy organised by source, date and importance. It includes a review of general renewable energy impact approaches, as well as a section on the costing of impacts. It is a search for common themes in the impact assessment of the different and disparate sources. It is a critical and fairly comprehensive review and survey of a selection of literature up to 1994 of environmental impacts of renewable energy sources. One of the main conclusions is the need for a common framework in comparison of 'apples and oranges'. Other major points are the lack of reliable methodologies for comparison, the inadequacy of current definitions and analyses of natural environments, and that sensitivity to impacts is likely to result from the fact that natural energy flows power the environment. A characteristic, the diffuse energy density of renewable energy sources, is pointed out.

Since the work was carried out there have been advances in some areas as reviewed above, though the author believes many of the issues remain unresolved.

A criticism of LCA techniques applied to renewable energy systems where complete cycles are required can be found in 'Life Cycle Assessment of

Energy from Biomass Waste Sources' (A. Clarke 2000). There is a description of an analysis and model of impacts from renewables as applied to biomass and 'green' criteria. The importance of soil carbon in determining CO_2 neutrality and 'sustainability' is emphasised. Another theme concerns the problems of employing combustion in energy conversion.

A version as a journal paper was produced – 'An Assessment of Biomass as an Energy Source: The Case of Energy from Waste' (Clarke and Elliott 2002). The point was made of the need for thorough examination of the carbon balance.

Some of the literature on the impacts of hydro electricity, the main case study focus of this book, is reviewed below, in addition to the reviews of key issues in the literature, assessment techniques, green electricity criteria, biomass and biofuel controversy, and the author's own works.

2.7 Hydro electric power

Hydro electric power (HEP) is the oldest, most mature renewable electricity-generating source and it has the largest installed capacity worldwide of the eight renewable energy sources that are used to generate electricity, as distinct from traditional biomass use for heating, cooking and lighting. There is as a result a considerable body of literature available on the subject of environmental effects and impacts from this source, and a selection is reviewed here. HEP has thus been selected as the most amenable source to use to investigate the hypothesis.

There has been considerable debate on the impacts of hydro, especially large hydro, with opposition growing since the very large developments in the period 1930–1970. Such large schemes are now generally confined to developing continents such as South East Asia, Africa and South America. HEP developments have tended to lose their good reputation in developed countries and active opposition networks exist – for example, the Wild Rivers Campaign in the USA (American Rivers 2009). Opposition has also mounted to large hydro in developing countries, based in part on their alleged environmental impact (McCully 1996, 2004).

The political nature of the controversy over impacts of HEP can be seen in a paper on HEP and the environment in the subcontinent, 'Hydropower and the Environment in India' (Ranganathan 2000). This is a pro-development debate/polemic which deplores adversarial decision-making processes.

The potential environmental problems have, of course, not gone unresearched. A major work on HEP impacts is that by Petts, 'Ecology of Impounded Rivers: Perspectives for Ecological Management' (Petts 1984). This is an important work on the effects of dams and reservoirs, including biochemical and biological issues, their geomorphology effects such as river and bed load, erosion, water stratification and ecology. Petts raises questions over the degree of change rivers experience when dammed, and emphasises the 'river continuum' concept whereby changes continue to have effects downstream. He proposes a classification scheme and structure for addressing impacts and includes much detailed evidence and data.

That the HEP industry is trying to address criticisms can be seen in a paper by Trussart *et al.* (2002), on the topic of environmental effects and solutions for HEP. Titled 'Hydropower Projects: A Review of Most Effective Mitigation Measures', it covers impact avoidance measures, mitigation measures, and compensation and enhancement measures (Trussart *et al.* 2002). A thorough checklist is provided, together with discussion. The authors consider that effective measures can be taken to reduce impacts, but make the point that effectiveness of measures is not well known due to lack of comprehensive monitoring.

Support for HEP in fast-developing countries can be seen in a paper on power generation in China, which defends China's HEP programme. 'Hydropower in China' by Li (2000) describes China's power situation, development and modernisation demands, the options available, fuel and pumped storage, and justifies the HEP programme. Li believes China's HEP policy is right and Western environmentalist criticisms are wrong.

The subject of social, environmental and economic changes as it applies to HEP development and an explanation of changes in HEP development is described in Oud (2002). This is a comprehensive account of past and present development and rationale. The author claims hydropower can adapt to sustainability requirements, economic liberalisation and increased social demands, and can compete economically. Methods by which HEP can adapt are included, e.g. design and flexibility features, such as multiple depth dam sluices.

The importance of HEP, hydropower developments worldwide and its potential in different regions of the world are the subject of a paper by Bartle 'Hydropower Potential and Development Activities' (Bartle 2002). This is a market survey of expansion in different countries asserting that hydropower is expanding especially in developing areas due to its maturity, economy, long life, and abundance of the resource. The author's view is that environmentally and socially well planned hydropower has a major role to play in future world energy supply, despite the 'recent wave of public opposition'.

The evolving environmental regulation of HEP can be observed in 'The Environmental Legal and Regulatory Frameworks: Assessing Fairness and Efficiency' by Berube and Cusson (2002). The paper is concerned with hydro statutory environmental regulatory assessment, the influence of environmental and strategic environmental assessments. This is a legislative analysis of policy, project planning, screening and scoping. The authors make the point that there is confusion in the process, and believe that hydro electric power has lost out. They point to the need for strategic environmental assessments.

Sustainable hydro electric development proposals are contained in 'Recommendations for Sustainable Hydroelectric Development' (Klimpt *et al.* 2002). This paper has a discussion and checklist for greener hydro procedures. The point is made that comparisons need to be made on a like-for-like basis; the benefits as well as impacts should be included in LCAs and environmental assessments. Klimpt's view is that hydropower has made much progress in the last 20 years and can be 'green'.

The research activities of the World Commission on Dams (WCD), a former industry association, on impacts of HEP are the subject of 'WCD Cross Check Survey Final Report' by C. Clarke. This is a survey of dams' performance, intended and actual developments, by the WCD (C. Clarke 2000).

However, the opposition has continued to press its case.

In 2004 the International Rivers Network (IRN) produced a declaration calling for large hydropower to be excluded as a renewable energy option in the UN-initiated Clean Development Mechanism, a green project support system for developing countries. The IRN claimed that large hydro projects have 'major social and ecological impacts', most obviously in terms of social dislocation resulting from the need to move people from the inundated areas, and said that efforts to mitigate local eco-impacts typically fail. Moreover, it said the alleged economic and social benefits of large hydro projects are often illusory ('Large hydro does not have the poverty reduction benefits of decentralised renewables'), and while it supported small hydro, it claimed that including large hydro in funding initiatives would 'crowd out funds for new renewables'.

2.7.1 HEP and GHG emissions

The IRN had also claimed that large reservoirs 'can emit significant amounts' of GHGs like methane, thus undermining hydro's claim to being 'climate friendly'. This is one conclusion from an earlier report produced by the WCD, 'Dams and Development: A New Framework for Decision-making' (WCD 2000b [2001b]). This looked at the emissions of GHGs from dam reservoirs due to the rotting of biomass trapped when the reservoir is filled and then brought downstream subsequently, and compared these emissions with those from other energy sources.

After two years of work, the WCD concluded that

> All large dams and natural lakes in the boreal and tropical regions that have been measured emit greenhouse gases . . . some values for gross emissions are extremely low, and may be ten times less than the thermal option. Yet in some cases the gross emissions can be considerable, and possibly greater than the thermal alternatives.

The WCD notes that

> the flooded biomass alone does not explain the observed gas emissions. Carbon is flowing into the reservoir from the entire basin upstream, and

other development and resource management activities in the basin can increase or decrease future carbon inputs to the reservoir.

In response, the International Hydropower Association (IHA), an industry non-governmental association, argued that the IRN's assertions about emissions were not properly backed up. Only a few studies had been done, and in any case, although HEP reservoirs can emit methane, what matters are the net emissions, i.e. the emissions with the dam in place compared with those that were generated from that area beforehand.

The IHA claimed that:

1 Tropical rivers, floodplains and wetlands, in their natural state, are large emitters of GHGs.
2 Natural lakes and rivers, in boreal and temperate climates, have emissions similar to reservoirs older than ten years.
3 Dams have little effect on overall emissions, because virtually all the carbon flowing in rivers tends to be emitted to the atmosphere and only a very small fraction will sediment permanently in the ocean.
4 In tropical forests, seasonal rain naturally floods huge areas that have similar ecological conditions to those of tropical reservoirs.
5 In boreal reservoirs, a few years after impoundment, aquatic productivity and water quality are similar to that in natural lakes.

The allegation that CO_2 emissions are increased by HEP reservoirs was also refuted by Gagnon in a paper titled 'The International Rivers Network Statement on GHG Emissions from Reservoirs, a Case of Misleading Science' (Gagnon 2002).

Gagnon states that HEP emissions of GHGs are probably no more than those of rivers and floodplains in a natural state. His view is that comparisons have not been made on the same basis, i.e. they have not assessed the net emission or increase due to the dam. He therefore proposes an improved methodology.

Richey states that carbon balances of tropical rivers may be neutral or even positive, in a paper on CO_2 emission and carbon transport of tropical rivers (Richey *et al.* 2002). His paper, 'Outgassing from Amazonian Rivers and Wetlands as a Large Tropical Source of Atmospheric CO_2' is a technical paper reporting on measurements and calculations of the carbon transport and CO_2 balances and emissions of Amazon Basin rivers. Richey states that linkages between land and water may be stronger than thought, with rivers representing a significant downstream translocation of carbon originally fixed by the forest.

On the same subject, an article (about a workshop meeting) titled 'Cleaning up Hydro's Act' (Newman 2000) notes that while LCAs have concluded that HEP can produce net emissions from reservoirs, so do natural lakes, and natural habitats may be either sources or sinks for carbon. Reservoirs cause 'a net change in GHGs from pre to post impoundment'. Boreal reservoir emissions were very low compared to tropical ones, but even there, the variation in GHG emissions from Brazilian HEP schemes ranged by a factor of up to 500.

However, a number of physical characteristics of good and bad sites for GHG emissions are given. Good sites have a high power density per area inundated, bad sites have low 'power density (in) kW m^{-2}'; deep reservoirs were found to be better than shallow ones in this respect, with canyon-like shapes better than dendritic (Newman 2000). Water residence time in the reservoir needed to be short rather than long. The size of the catchment area should be small rather than large. Temperature for a good site was cold and for bad sites warm. Nutrient and carbon input was low for good sites and high for bad sites.

However, simplistic impressions needed to be avoided since a reservoir will not necessarily emit more GHGs than a river before impoundment. The authors considered that more research was needed on the issues of pre-dam GHG emissions and sinks from natural habitats and the basin, as well as assessment of GHGs released from water passing through the dam, and observed GHG emission from the reservoir surface and inflows of carbon from upstream.

Some of the parameters given above for good and bad HEP sites associated with water quality can be linked to changes to the energy flow and have been used in this study (see Chapter 5).

2.8 Land use and diffuseness

A persistent criticism of renewable energy's practicality by nuclear and fossil fuel industries has been that the land use, i.e. area required, is too great; that is renewable sources are too diffuse (Walker 1995a). However, until fairly recently exactly *how* diffuse or concentrated sources are has not been used as a means of characterising the different sources.

Walker points to the importance of land use for renewable energy in a paper on planning issues, 'Energy, Land Use and Renewables' (Walker 1995a). He states that changes in energy policy have implications for land use, and that 'interrelated themes of land requirements, environmental impacts and public opposition, and planning policy are identified'. He notes the diffuse nature of renewable energy, stating that land availability is a potential constraint on renewable technologies.

Land use is raised as an issue by Gagnon *et al.* in a paper 'Life Cycle Assessment of Electricity Generation Options: The Status of Research in Year 2001', (Gagnon *et al.* 2002). Gagnon *et al.* compare other electricity options with hydro. They claim that hydro has the lowest land use per TWh. Their view is that shortcomings and omissions of assessments have militated against hydro. This work is discussed further in Chapter 3.

Caetana de Souza (2007) makes the point that modern HEP plants use much less land per unit of dependable generation than older HEP schemes, in a paper titled 'Assessment of Statistics of Brazilian Hydroelectric Power Plants: Dam Areas Versus Installed and Firm Power' (Caetana de Souza, 2007). His HEP assessment technique involves an index system of environmental impact assessment based on land area compared to installed capacity and firm power in Brazil. A thorough survey of land areas and installed capacity and capacity factors was carried out, and arranged in an index by region for Brazil as criteria for environmental impact assessment. The index is based on land area compared with installed capacity. Caetana de Souza's view is that capacity factors can be a useful parameter for impact assessment, and that more modern development can reduce land take. Modern developments can have a higher environmental index based on firm power/land area. The author also suggests that electricity generation per unit of land area would be a useful parameter, and that calculated volume of reservoirs could be another basis for environmental analysis.

The issue of land use is considered by MacKay (2008a), but in terms of the average power in watts per square metre, for renewable energy sources across the whole UK. MacKay (2008b) assesses the renewable resources available to the UK in a wide-ranging tour of energy options available for the UK, in a book *Sustainable Energy: Without the Hot Air.* He poses the question of 'whether a country like the United Kingdom, famously well endowed with wind, wave, and tidal resources, could live on its own renewables'.

He assesses the available energy resources in comparison with the current energy demand, from all sources and not just for electricity. He states 'all the available renewables are diffuse – they have small power density' (MacKay 2008a: 181). For this reason he considers that it will be difficult for bulk renewable energy supply not to be large and intrusive. The method of determining the power flux density is to take the average resource on a country-wide basis and this is then turned into W m^{-2} and totalled as kWh m^{-2} per inhabitant per day. Certainly it is true to state that the power density (the measure of intensity/diffuseness) is small on this basis.

However, whether like is really being comparing with like could be questioned, as for example, some of the space comparisons are for water areas, or the sea or estuaries, and some are land areas. This is acknowledged by the author, as is the fact that some of these land areas will be available for several renewable sources simultaneously. For example, for solar energy, panels allow no other primary renewable energy land use to take place below them at all (apart from geothermal), i.e. assuming the occupation in terms of interception of light is 100 per cent; but that for wind energy allows normal farming practices, either grazing or arable, to continue. There could, however, be solar panels on the ground occupying the same space as a windfarm, apart from the actual towers themselves. In fact, the wind turbines should be described as being spread *across* the land area. That same land area is, of course, also acting as a catchment area for rivers which may be harnessed for HEP.

However, MacKay does try to make a practical assessment of the different resources by incorporating some of the constraints of each renewable source in terms of collector characteristics, such as counting only south-facing roofs for solar water heaters or photovoltaic panels.

His estimations for HEP land use are presumably averaged areas for the reservoirs, as opposed to 'run of the river' or schemes with leats running along a valley side.

MacKay concludes that 'it *would* be possible for the average European energy consumption of 125 kWh/day per person to be provided from these country-sized renewable sources' (MacKay 2008b: 3), provided that economic constraints and public objections can be set aside. However, he questions whether the scale of development would actually be acceptable and has therefore worked up a range of scenarios or energy plans involving substantial reduction of energy demand and different permutations of renewable sources, 'clean coal' (with carbon capture and sequestration) and nuclear power.

Smil (2015), considers power density, defined as W m^{-2} as a prime factor in energy source feasibility assessment, including renewable energy, though he uses this parameter mainly in the context of land use, i.e. footprint. He states that current use of fossil and uranium energy resources have power densities, in terms of land use, two to five orders of magnitude greater than renewable sources. He makes the point that current energy systems mainly exploit concentrated energy sources and distribute them to more diffuse uses, while future energy systems based on renewable sources would do the reverse, exploiting diffuse sources and concentrating them for more power-dense applications. This fundamental shift would emphasise the constraint of use of land area by low power flux density renewable sources. Like MacKay, Smil applies power density assessment to determine resource size and feasibility.

2.8.1 Extraction of energy proportion from flows, in the literature

The concept of abstraction of the kinetic energy of natural energy flow, which is a key issue considered in this book, is pursued by Dacre in 'A Scoping Study to Establish a Research Programme on the Environmental Impacts of Tidal Stream Energy' (Dacre 2002), a scoping study and generic environmental impact approach to tidal stream energy. This is a fairly thorough analysis, using some similar methodology to that used in this study. Among the conclusions and recommendations, the study identified some key environmental issues for tidal current energy development; in particular, 'the impact of extracting energy and the effects on tidal flow patterns, sedimentation processes and sea bed morphology'. This, among other areas, was highlighted as needing further knowledge. The development of a parametric model was another research area requiring further work.

The Pentland Firth tidal current energy project's environmental impact is the subject of the extended Executive Summary report 'Pentland Firth Tidal

Current Energy Feasibility Study – Phase 1', by Bryden *et al.* of Robert Gordon University and Scottish Enterprise Ltd (Bryden *et al.* 2002). This impact assessment on the Pentland Firth tidal current energy scheme identifies the likely impacts on the physical and ecological environment which include 'possible changes in tidal patterns and wave climate through structure presence and through potential energy reduction and vorticity effects', as well as 'changes in water turbidity and quality from seabed disturbance, gearbox leakages etc' (Bryden *et al.* 2002: 16). Other impacts are the disturbance to wildlife such as sea birds from cable laying and installation of the turbines, as well as noise and wildlife collision risk. Among the recommendations for further research on physical and ecological issues is '[f]urther extensive research and hydrodynamic modelling . . . to quantify the scale of likely effects in which devices may affect current patterns and velocities and thus local sediment movement' (18). Overall the report's authors consider that impacts from tidal current energy devices are only moderate, despite some potential disruption to the environment.

A mention of the concept of 'proportion', a key factor explored in this book, is also made by Charlier in 'Sustainable Co-Generation from the Tides: A Review' (Charlier 2003). He cites researchers suggesting that a 'lower proportion of extraction may reduce impacts'. This is a comprehensive review of tidal power around the world, including Russian and Chinese plants. The author states that tidal energy works, needs no new technology and is environmentally more benign than fossil or nuclear alternatives. Capital costs of tidal remain high, but it has long life times, and low running costs are possible.

Related to the methodology used here and the concept of 'proportion', the cumulative effect of impact of wind turbines is to be taken into account in updated government planning policy guidelines for renewable energy sources. 'Planning Policy Statement 22: Renewable Energy' (HMSO 2004) itemises modified rules on planning policies as a result of about ten years' experience in facilitating new government renewable energy targets. A prescriptive (restricting and directing) approach to renewable energy planning is disallowed; for instance, no buffer zones outside designated areas are allowed (but the effect on such zones is considered material). Also, no urban (or other) prescriptive prohibition was allowed, but developments appraised are to be undertaken on their merits and impacts, on a case-by-case basis. No substitution of offshore capacity for onshore was allowed either.

2.8.2 Macro environmental impacts from large-scale use of renewable energy

As the deployment of renewable energy has proceeded apace in the twenty-first century, various studies have considered the potential effects of very large-scale use of renewable energy on the earth's thermodynamic energy and climatic system. Such studies range from the issue of renewable energy's place in the overall global energy balance and thermodynamic system, e.g. 'How Does the Earth System Generate and Maintain Thermodynamic Disequilibrium and What Does It Imply for the Future of the Planet?' (Kleidon 2012), 'Estimating Global Maximum Surface Wind Power Extractability and Associated Climatic Consequences' (Miller *et al.* 2011) and 'Thermodynamic Limits of Hydrologic Cycling within the Earth System: Concepts, Estimates and Implications' (Kleidon and Renner 2013), to more detailed studies considering impacts from very large-scale use of individual technologies such as wind and solar. Examples of such papers are 'Can Large Wind Farms Affect Local Meteorology?' (Baidya Roy and Pacala 2004), 'Investigating the Effect of Large Wind Farms on Energy in the Atmosphere' (Sta Maria and Jacobson, 2009), 'Saturation Windpower Potential and Its Implications for Wind Energy' (Jacobson and Archer 2012) or 'Regional Climate Consequences of Large-Scale Cool Roof and Photovoltaic Array Deployment' (Millstein and Menon 2011) and 'Impacts of Array Configuration on Land-Use Requirements for Large-Scale Photovoltaic Deployment in the United States' (Denholm and Margolis 2008). In general, such studies attempt to quantify the maximum limits to extraction of renewable energy flows and predict effects through thermodynamic theory and changes in entropy. Models of atmospheric, oceanic and geophysical energy flows have been used to estimate effects on large-scale geo-energetic processes and so set limits to exploitation. Kleidon (2012) traces his thermodynamic disequilibrium approach to that of Lovelock (1975). Kleidon uses the concept of 'free energy' to propose that only a small proportion of the total energy received from the sun is available as 'free energy'. He states that although the sun sends large amounts of energy to earth, i.e. about 1.2×10^{17} W, or 120,000 TW – roughly 8,000–10,000 times human energy use – most of this is unavailable since it powers natural processes such as the hydrological cycle and the atmospheric circulation system. 'Free energy' is defined as that energy available after energy to create disequilibrium in order to power the system has been accounted for. Kleidon points out that 'Generating disequilibrium in a thermodynamic variable requires the extraction of power from another thermodynamic gradient, and the second law of thermodynamics imposes fundamental limits on how much power can be extracted.' While in general terms this must be true, Kleidon quantifies the 'free' energy available for wind energy at just a few tens of TW, 18–34 TW, in Miller *et al.* (2011), 'notably less than recent estimates that claim abundant wind power availability'. However, there can be the appearance of 'knocking down a straw man' in some of these works, with, for example, allusion to claims of 'limitless' amounts of renewable energy. Some studies exhibit this tendency by using a hypothetical scenario, e.g. that of powering all or a very large proportion of the global energy demand from wind energy, which though useful as a 'thought experiment', has never been seriously contemplated. Similar papers consider the HEP resource in equivalent terms (Kleidon and Renner 2013). The actual estimates of Kleidon's 'free energy' have been criticised as too low by, for example Jacobson and others. However, Kleidon's adopted approach considers natural energy conversion processes in terms of entropy, the second law of thermodynamics, and 'thermodynamic disequilibrium', which has merit on a macro-scale, though in practice it may represent an over-simplification of the processes involved. Miller *et al.* state that a thermodynamic disequilibrium approach rather than an 'engineering type' approach should be adopted for estimating ultimate limits to renewable energy exploitation. While such an approach may be useful in some respects, e.g. in development of theory, an engineering approach could be considered to be the practice necessary to test the theory. Additionally, models of global energy circulation systems can only ever be idealised simplified versions of reality. In practice, if a decline in wind speeds over large areas were to be experienced as a result of large-scale wind energy exploitation, no further wind turbines would be installed there. An engineering approach would be sensitive to evidence of depletion, since it would be uneconomic to be otherwise. The modular nature of much renewable energy technology could thus avoid large-scale depletion from over exploitation.

Sta Maria and Jacobson (2009) come to different conclusions regarding the potential for wind energy, in a study which estimates energy loss to be only 0.007 per cent in the lower 1 km of the atmosphere if all of the world's electricity were supplied by wind energy. Jacobson and Archer (2012) put the ultimate limits for wind energy at >250 TW at 100 m height globally and 80 TW at 100 m over land and coastal areas outside of Antarctica, and 380 TW from jet streams at 10 km altitude. They state that there is no fundamental barrier to obtaining half of the world's energy or 5.75 TW from wind energy, or several times electricity production, in a paper titled 'Saturation Wind Power Potential and Its Implications for Wind Energy'. Jacobson and Archer (2012) outline their method of determining the maximum theoretical wind power potential on earth, which is the point at which additional turbines do not generate additional electricity – in total. This they term the 'saturation wind power potential' (SWPP). The authors use a physical model of energy extraction from the atmosphere, taking into account the 59 per cent Betz limit, extraction from 100 m height as this is typically the hub height of current wind turbines, as well as energy conservation, conservation of heat energy from turbulence generated due to surface roughness, and conservation of energy due to electricity use (i.e. the resulting heat which is returned to the atmosphere, with a small part contributing to regenerated kinetic energy). The thoroughness of the modelling method extends to use of six different atmospheric layers from which energy is extracted and application to numerous windfarms worldwide simultaneously, as well as momentum extraction feeding back to global dynamics rather than a limited regional area.

This physical approach appears to be more realistic and soundly based than purely top-down modelling approaches, as it incorporates more of the actual features of wind energy extraction technology, e.g. the Betz limit, which limits the proportion that can be extracted from an extended fluid flow to 59 per cent, as well as conservation of energy.

Nevertheless, in various respects, Kleidon comes to fairly reasonable conclusions in that from the perspective of the earth's energy systems, the harnessing of very large amounts of energy should be the first 'port of call' at a low entropy level, where in fact the greatest resources exist, i.e. solar, as opposed to the forms of subsequent conversion, e.g. wind, wave and HEP. Kleidon also points to the conversion of solar radiation into chemical form in plants through photosynthesis as a means of retaining the low entropy of the energy form and he suggests means to increase the total global photosynthetic conversion.

At a more intermediate level, somewhere between thermodynamic disequilibrium theory and an engineering approach, work has been carried out on the more regional macro effects of large areas of solar parks, and very large windfarms on local meteorological conditions, e.g. Millstein and Menon (2011) and Baidya Roy and Pacala (2004). In particular, the effects of very large solar parks and very large windfarms on convection currents from solar radiation ground warming are considered, as well as the increased albedo (reflectivity) from solar PV arrays and the extent of energy extraction. An increase in albedo could result in reduced heating of the ground surface, leading to reduced air convection, and thus to a cooling effect. This might become regionally significant on a large scale and could influence cloud formation, reducing it by a small proportion. Currently PV commercial conversion efficiency is unlikely to be much more than about 20 per cent, which as an energy sink may not prove very significant, though changes in albedo may have more effect. Additionally, panels will tend to shelter the ground surface from winds, altering near-surface air flows to some extent.

Baidya Roy and Pacala (2004) state that very large windfarms may alter convection currents by causing increased surface roughness and thus increased friction, wake-induced turbulence, as well as slowing the near-surface wind speeds through extraction of energy. Wang and Prinn (2010) state that if wind energy were to supply 10 per cent of energy in 2100, this could result in surface warming exceeding 1 °C over land installations and 1 °C cooling over sea installations, in a study using a three-dimensional model to simulate potential climate effects. Significant warming or cooling remote from land or water installations was forecast from the model, though they state that the validity of their method of extrapolation needs further study. The regional climatic and radiative effects of modifying surface albedo to mimic massive deployment of cool surfaces such as roofs and pavements, and photovoltaic arrays across the USA, were investigated by Millstein and Menon (2011). They used a fully coupled regional climate model (WRF) to investigate surface albedo change effects on surface temperature, precipitation and average cloud cover. Hypothetical 1 TW_p solar arrays in the Mojave Desert, California were modelled by darkening of the surface albedo, resulting in local afternoon temperature increases of +0.4 °C, which also affected local and regional wind patterns in a 300 km radius. The authors state that there were lower magnitude but statistically significant changes to temperature and radiation across the domain, though there was no 'significant change to summertime outgoing radiation, when averaged over the full domain as interannual variation obscured more local forcing' (Millstein and Menon 2011).

It is clear from such studies that very large deployment of renewable energy could potentially have impacts which could be significant locally and regionally, and possibly at a greater scale, though these effects are as yet uncertain in scale and significance. Disciplines such as thermodynamic disequilibrium are not yet fully established with sufficient development and proof of evidence, and so conclusions remain somewhat hypothetical. Forecasting impacts from large-scale renewable energy by models of atmospheric circulation can be useful to test potential effects as well as the parameters, though as yet these appear relatively small, and again uncertain. As Kleidon points out, global energy transfer processes are competitive – if one form, e.g. conveyance of heat energy by moving air, is resisted or blocked, then another will tend to take its place, e.g. oceanic heat conveyance. The complexity of the global atmospheric, oceanic and earth energetic systems involving many different coupled processes makes forecasting of effects from changes in conditions difficult and uncertain. There is the danger of double accounting or omitting processes. For example, biomatter cultivated for food or fuel will still perform many or at least some of the transpiration roles with attendant atmospheric functions that more 'natural' flora, e.g. forests, perform. Large grassland areas of the land surface are also a natural part of the earth's terrain – without interference from humans. Similarly, the friction generated by the surface roughness of forested areas can be equated to that of some wind energy exploitation.

The potential capacity of renewable energy is of course not unlimited, but finite and for human purposes unending. While most of the natural energy flows through earth will need to continue performing the natural functions, human energy needs currently at about 17 TW, even expanded, can probably be satisfied by using a small proportion of the natural energy flows without too much impact – if intelligent design parameters are taken into account.

However, such studies, particularly those of local and regional effects of large-scale renewable energy deployment, use several of the key parameters for assessment proposed in this book, i.e. power flux density, proportion intercepted, efficiency of conversion and proportion extracted.

2.9 Conclusion

Although much literature exists on the subject of the environmental impacts from renewable energy, and this has been a developing field, there have been few systematic physical approaches that can allow comparison between and assessment of all the different renewable energy sources or provide a template for the assessment of possible future sources.

More progress appears to have been made on developing approaches that can unify the valuation of environmental effects into a generalised measure, e.g. cost, such as in the ExternE project, or on social decision-making techniques such as MCA or multi-attribute decision making (MADM).

Due to the huge variation in characteristics of different sources, most studies concentrate on describing and modelling the environmental effects and impacts of individual sources and technologies. Most studies are concerned with itemising the various effects and then suggesting how they might be minimised. The result can be a check list with a range of potential actions taken in response. The disparity between the varied renewable sources and their impacts has been termed comparing 'apples and pears' (Holdren *et al.* 1980) and has been wide enough to forestall attempts to establish a common framework.

The response to this has been to compare the environmental damage from disparate sources and put a cost on it, as has been done by the ExternE project, which uses an amalgamation of cost–benefit analysis and LCA. The latter may not be wholly appropriate for renewable energy systems, which utilise flows of energy, rather than products or processes having a definite beginning and end stage. The Fuel Cycle approach adopted is therefore considered by this author to be less appropriate to most forms of renewable energy.

With regard to HEP, the review of key works in this area shows that the industry is well aware of environmental criticisms and is addressing them to an extent. The importance of HEP in world electricity generation, and its establishment as a major industry, is to some extent used to justify impacts, for instance on the basis of kWh per unit of impact, with claims made that this is lower than with other sources. This issue is pursued further in Chapter 3, and in more detail in Chapters 5–7.

While some common themes emerge from the literature on environmental impacts of renewable energy sources, these tend to be in the areas of minimum standards – e.g. noise, air and water pollution – connected to planning regulations. Alternatively issues of carbon emissions saved or carbon neutrality emerge, for example in respect to biomass or HEP. Common themes can be found in the approach comparing economic costs of impacts, as adopted by the ExternE project, although as the range of costs cited can be very wide, much uncertainty exists with this approach.

One significant common theme that does emerge is the topic of land use and hence the diffuseness of renewable energy sources. Few works have adequately quantified the diffuseness in power flux density terms, i.e. kW m^{-2}, and linked this to impacts, apart from MacKay and Smil in terms of land use and overall resource size.

There are, in the literature, some references to the concept of energy extraction from a flow, for example Holdren *et al.* (1980), Dacre (2002) and

Bryden *et al.* (2002), and to the proportion extracted, especially latterly in the context of work on the impacts and constraints of developing tidal stream energy. It appears that quantification of these parameters is not yet complete. In part this is due to the need to model the baseline, i.e. pre-existing energy flow, before development of the renewable energy scheme modifies it. As Bryden *et al.* point out, this is in itself a considerable challenge (Bryden *et al.* 2002).

That natural energy flows power the natural environment is a point made by Holdren *et al.*, and this becomes a significant factor when 'large enough interventions in these natural energy flows and stocks' are made. This implies that proportion of energy extracted might be a significant parameter.

In 1990, the Watt Committee noted that no assessment methods they had seen work very well and consistently (Watt Committee 1990). This still seems to be the case. This issue is pursued further in the next and following chapters.

More recently in the twenty-first century, studies have been carried out on the macro environmental impacts of renewable energy to determine ultimate limits to exploitation and forecast effects. Studies of macro environmental impacts from very large-scale use of renewable energy deployment have been carried out using modelling approaches and indicate possible limits to overall exploitation of certain renewable energy sources, though these are at present relatively uncertain. Potentially significant, though relatively small, impacts on local and regional scales have been estimated, but results are again uncertain due to a lack of substantiated evidence as yet.

A variety of approaches have been adopted, top-down or bottom-up, using atmospheric, oceanic and land models, resulting in a variety of different conclusions. These range from a limited renewable energy resource potential, e.g. 18–34 TW total wind energy resource (Kleidon 2012), with significant environmental effects, to much more substantial resource availability, e.g. 80 TW for onshore and coastal wind energy and relatively minor environmental effects (Jacobson and Archer 2012), using a more physically based model.

Although there is considerable uncertainty about these approaches and results, the methods used by such studies do tend to employ some of the parameters identified in this book, i.e. power flux density, proportion intercepted, efficiency of conversion and proportion extracted.

Bibliography

Abbasi, S.A. and Abbasi, N. (2000) 'The Likely Adverse Environmental Impacts of Renewable Energy Sources', *Applied Energy*, Vol. 65 No. 1–4: 121–44.

American Rivers (2009) Website, online resources: www.americanrivers.org/site/PageServer, accessed 19 January 2009.

Baidya Roy, S. and Pacala, S.W. (2004) 'Can Large Wind Farms Affect Local Meteorology?', *Journal of Geophysical Research*, Vol. 109, doi:10.1029/2004JD004763.

Bartle, A. (2002) 'Hydropower Potential and Development Activities', *Energy Policy*, Vol. 30: 1231–9.

Bell, D., Gray T. and Haggett, C. (2005) 'The "Social Gap" in Wind Farm Policy Siting Decisions: Explanations and Policy Responses', *Environmental Politics*, Vol. 14 No. 4: 460–77.

BERR (2008) 'Digest of UK Energy Statistics', BERR, Department of Business Enterprise and Regulatory Reform, online resources: http://stats.berr.gov.uk/energystats/dukes5_5.xls, accessed 20 November 2008.

Berube, G. and Cusson, C. (2002) 'The Environmental Legal and Regulatory Frameworks: Assessing Fairness and Efficiency'. *Energy Policy*, Vol. 30: 1291–8.

Boyle, G. (ed.) (1996) *Renewable Energy Power for a Sustainable Future,* Oxford University Press in conjunction with the Open University, Milton Keynes.

Boyle, G. (ed.) (2004) *Renewable Energy: Power for a Sustainable Future* (2nd edition), Oxford University Press in conjunction with the Open University, Oxford and Milton Keynes.

Bratrich, C. and Truffer, B. (1998) *Green Power Publications*, EAWAG Swiss Federal Institute for Aquatic Environmental Science and Technology, Switzerland.

Bratrich, C., *et al.* (2000) 'A Multidisciplinary Project to Reduce Environmental Impact of Hydropower Generation', EAWAG Swiss Federal Institute for Aquatic Environmental Science and Technology, online resources, www.oekostrom.eawag.ch/veroeffentlichungen/ecohydraulics_bratrich.pdf 999/2000, accessed 7 January 2009.

Bryden, I., Dacre, S. and Bullen, C. (2002) 'Pentland Firth Tidal Current Energy Feasibility Study: Phase 1', Report Extended Executive Summary 2002 March, Robert Gordon University/Scottish Enterprise Ltd.

Bucholz, T., Rametsteiner, E., Volk, T. and Luzardis, V. (2009) 'Multi Criteria Analysis for Bioenergy Systems Assessments', *Energy Policy*, Vol. 37: 484–95.

Caetana de Souza, A. (2007) 'Assessment of Statistics of Brazilian Hydroelectric Power Plants: Dam Areas Versus Installed and Firm Power'. *Renewable and Sustainable Energy Reviews*, Vol. 12: 1843–63.

Chakrabortty, A. (2008) 'Exclusive We Publish the Report They Didn't Want You to Read', *Guardian*, 10 July, online resources: www.theguardian.com/environment/blog/2008/jul/10/exclusivethebiofuelsreport, accessed 16 September 2015.

Charlier, R.H. (2003) 'Sustainable Co-Generation from the Tides: A Review', *Renewable and Sustainable Energy Reviews*, Vol. 7: 187–213.

Clarke, A. (1993) 'Comparing the Impacts of Renewables: A Preliminary Analysis', TPG Occasional Paper Technology Policy Group 23, Open University, Milton Keynes.

Clarke, A. (1994) 'Comparing the Impacts of Renewables', *International Journal of Ambient Energy*, Vol. 15 No. 2.

Clarke, A. (1995) 'Environmental Impacts of Renewable Energy: A Literature Review,' Thesis for a Bachelor of Philosophy Degree, Technology Policy Group, Open University, Milton Keynes.

Clarke, A. (1998) 'Buyer Beware', *New Scientist*, 13 June.

Clarke, A. (2000) 'Life Cycle Assessment of Energy from Biomass Waste Sources', Energy and Environment Research Unit, Report 075, Open University, Milton Keynes.

Clarke, A. and Elliott, D. (2002) 'An Assessment of Biomass as an Energy Source: The Case of Energy from Waste', *Energy and Environment Journal*, Vol. 13: 27.

Clarke, A.D. (1988) 'Windfarm Location and Environmental Impact', Charter for Renewable Energy and Network for Alternative Technology and Technology

Assessment, c/o Energy & Environment Research Unit, Open University, Milton Keynes.

Clarke, C. (2000) 'WCD Cross Check Survey Final Report', World Commission on Dams, online resources: www.dams.org/kbase/survey, accessed 20 July 2009; www.dams.org/docs/kbase/survey/ccsmain.pdf, accessed November 2000.

Cross, N. (1982) 'Methods Guide', Control of Technology, Unit 8, prepared for Course Team by Nigel Cross, Open University Press, Milton Keynes.

Dacre, S. (2002) 'A Scoping Study to Establish a Research Programme on the Environmental Impacts of Tidal Stream Energy', Department of Trade and Industry, Contractor Robert Gordon University Aberdeen, Energy Technology Support Unit, T/04/00213/00/REP.

DECC (2009a) *The UK Low Carbon Transition Plan*, Department of Energy and Climate Change, London.

DECC (2009b) *Offshore Energy SEA Environmental Report*, Department of Energy and Climate Change, London.

Denholm, P. and Margolis, R.M. (2008) 'Impacts of Array Configuration on Land-Use Requirements for Large-Scale Photovoltaic Deployment in the United States', paper presented at SOLAR 2008 – American Solar Energy Society (ASES), San Diego, California, 3–8 May.

Devine-Wright, P. and Devine-Wright, H. (2006) 'Social Representations of Intermittency and the Shaping of Public Support for Wind Energy in the UK', *International Journal of Global Energy Issues*, special issue on intermittency, Vol. 25 No. 2/4: 243–56.

EEA (1998) 'Life Cycle Assessment: LCA A Guide to Approaches, Experiences and Information Services', Report no. 6, EEA European Environment Agency, online resources: http://reports.eea.europa.eu/GH-07-97-595-EN-C/en/Issue%20report%20No%206.pdf, accessed 13 January 2009.

Elliott, D. (1994) 'Public Reactions to Wind farms: The Dynamics of Opinion Formation', *Energy and Environment*, Vol. 5 No. 4: 343–62.

Elliott, D. (2003) *Energy Society and Environment*, London, Routledge.

Ellis, G., Barry, J. and Robin, C. (2006) 'Renewable Energy and Discourses of Objection Towards Deliberative Policy Making: Summary of Main Research Findings', online resources: www.qub.ac.uk/research-centres/REDOWelcome, accessed 20 July 2009.

EREC (2007) 'Renewable Energy Target for Europe: 20% by 2020', European Renewable Energy Commission, online resources: www.erec.org/fileadmin/erec_docs/Documents/Publications/EREC_Targets_2020_def.pdf, accessed 19 January 2009.

EU (2003) 'EU Biofuels Directive, Directive 2003/30/3 of the European Parliament and the Council of 8th May 2003 on the Promotion of the Use of Biofuels or Other Renewable Fuels for Transport', *Official Journal of the European Union European Union*, online resources: http://ec.europa.eu/energy/res/legislation/doc/biofuels/en_final.pdf, accessed 15 April 2009.

Europa (2008) 'Memo on the Renewable Energy and Climate Change Package', Europa Press Release Rapid, online resources: http://europa.eu/rapid/pressReleasesAction.do?reference—EMO/08/33&format=HTML&aged=0&language=EN&guiLanguage=en accessed 19 January 2009.

EWEA (2008) 'Wind Energy Statistics Windmap', European Wind Energy Association, online resources: www.ewea.org/fileadmin/ewea_documents/mailing/windmap-08g.pdf, accessed 30 December 2008.

EWEA (2013), 'EWEA Annual Report 2013', European Wind Energy Association, online resources: www.ewea.org/publications/reports/ewea-annual-report-2013, accessed 14 October 2015.

ExternE (1995a) *Externalities of Energy*, Vol. 1, European Commission, Luxembourg.

ExternE (1995b) *Externalities of Energy*, Vol. 6, European Commission, Luxembourg.

FoE (2008), 'EU Biofuel Targets Must Go', online resources: www.foeeurope.org/press/2008/Jul07_EU_biofuel_targets_must_go.html, accessed 25 January 2016.

FoE (2010) '"Sustainable" Palm Oil Driving Deforestation: Biofuel Crops, Indirect Land Use Change and Emissions', Briefing Note, Friends of the Earth. Online resources: www.foe.co.uk/sites/default/files/downloads/iluc_palm_oil.pdf accessed 16 September 2015.

Fridley, D. (2010), 'Nine Challenges of Alternative Energy', in Heinburg, R. and Lerch, D. eds, *The Post Carbon Reader: Managing the 21st Century's Sustainability Crises*, Watershed Media, Healdsburg, CA.

Gagnon, L. (2002) 'The International Rivers Network Statement on GHG Emissions from Reservoirs, a Case of Misleading Science', International Hydropower Association, online resources: http://philip.inpa.gov.br/publ_livres/Other%20side-outro%20lado/Hydroelectric%20emissions/Gagnon%20IHA%202002.pdf, accessed 31 December 2008.

Gagnon, L., Belanger, C., Uchiyama, Y. (2002) 'Life Cycle Assessment of Electricity Generation Options: The Status of Research in Year 2001', *Energy Policy*, Vol. 30: 1267–78.

Gallagher, E. (2008) 'The Gallagher Review of the Indirect Effects of Biofuels Production', Renewables Fuels Agency, online resources: www.renewablefuels agency.org/_db/_documents/Report_of_the_Gallagher_review.pdf, accessed 31 December 2008.

Hertwich, E.G., Gibon, T., Bouman, E., Arvesan, A., Suh, S., Heath, G., Bergesen, J., Ramirez, A., Vega, M. and Shi, L. (2015) 'Integrated Life-Cycle Assessment of Electricity-Supply Scenarios Confirms Global Environmental Benefit of Low-Carbon Technologies.' *PNAS*, Vol. 112 No. 20: 6277–82.

Hill, R. and Twidell, J. (1994) 'The Environmental Impacts of Energy Technology', UK Solar Energy Society, Conference C62 University of Westminster, London 11 May.

HMSO (2004) *Planning Policy Statement 22: Renewable Energy*, Her Majesty's Stationery Office, London.

Holdren, J.P., Morris, G. and Mintzer, I. (1980) 'Environmental Aspects of Renewable Energy Sources', *Annual Review of Energy*, Vol. 5: 241–9.

Homer-Dixon, T. (2007) *The Upside of Down: Catastrophe, Creativity and the Renewal of Civilisation*, Island Press, Washington, DC.

Humphreys, D. (2013), 'Low Carbon Energy Technologies', in P. Ranguram *et al.*, eds, *Environment: Sharing a Dynamic Planet, Carbon, Food, Consolidation*, Open University, Milton Keynes.

IEA (2010) *Energy Technology Perspectives 2010: Scenarios and Strategies to 2050*, OECD/International Energy Agency, Paris.

IPCC (2007) 'Climate Change 2007: Synthesis Report Summary for Policymakers', Intergovernmental Panel on Climate Change, online resources: www.ipcc.ch/pdf/assessment-report/ar4/syr/ar4_syr_spm.pdf, accessed 20 July 2009.

ISO (2006) 'Environmental Management: Life Cycle Assessment, Life Cycle Impact Assessment', online resources: www.iso.ch/iso/iso_catalogue/catalogue_tc/catalogue_detail.htm?csnumber=23153, accessed 19 January 2009.

Jacobson, M.Z. and Archer, C.L. (2012) 'Saturation Wind Power Potential and Its Implications for Wind Energy', *PNAS*, Vol. 109 No. 39: 15679–84.

Jensen, A. (1997) *Life Cycle Assessment*, European Environment Agency, Copenhagen.

Kleidon, A. and Renner, M. (2013), 'Thermodynamic Limits of Hydrologic Cycling within the Earth System: Concepts, Estimates and Implications', *Hydrology and Earth System Sciences*, Vol. 10: 3187–236.

Kleidon, A. (2012), 'How Does the Earth System Generate and Maintain Thermodynamic Disequilibrium and What Does It Imply for the Future of the Planet?' *Philosophical Transactions*, Vol. 370: 1012–40.

Klimpt, J., Rivero C., Puranen, H. and Koch, F. (2002) 'Recommendations for Sustainable Hydroelectric Development', *Energy Policy*, Vol. 30: 1305–12.

Krewitt, W. (2002) 'External Costs of Energy: Do the Answers Match the Questions? Looking Back at Ten Years of ExternE', *Energy Policy*, Vol. 30 No. 10: 839–49.

Li, F. (2000) 'Hydropower in China', *Energy Policy*, November.

Liebreich, M. (2008) 'Food Price Increases: Is It Fair to Blame Biofuels?' *New Energy Finance*, online resources: www.newenergymatters.com/docs/Press/NEF_RN_2008-05-27_Food_Price_Drivers.pdf, accessed 12/08/09.

Lovelock, J. E. (1975). *Thermodynamics and the recognition of alien biospheres*, Proc. Roy. Soc. Lond. B 189: 167–181.

Lovelock, J. and Epton, S. (1975). 'The Quest for Gaia'. *New Scientist*, Vol. 65: 304.

MacKay, D. (2008a) *Sustainable Energy without the Hot Air*, Cambridge UIT, online resources: www.inference.phy.cam.ac.uk/sustainable/book/tex/cft.pdf.

MacKay, D. (2008b) 'Technical Chapters to Sustainable Energy', online resources, http://without the hot air.com, accessed 20 July 2009.

McCully, P. (1996) *Silenced Rivers: The Ecology and Politics of Large Dams*, Zed Books, London.

McCully, P. (2004), 'Tropical hydropower is a significant source of greenhouse gas emissions: a response to the International Hydropower Association', International Rivers Network 2004, online resources: https://www.internationalrivers.org/files/attached-files/tropicalhydro.12.08.04.pdf accessed 8th June 2016

Mila i Canals, L., Clift, R., Cowell, S., *et al.* (2006) 'Definition of Best Indicators for Land Use Impacts for Life Cycle Assessment: Expert Workshop', 12–13 June, University of Surrey IAS, online resources: www.ias.surrey.ac.uk/workshops/DEFNBEST/cfp.php, accessed 13 January 2009.

Miller, L.M., Gans, F. and Kleidon, A., 2011, 'Estimating Maximum Global Surface Wind Power Extractability and Associated Climatic Consequences' *Earth System Dynamics*, online resources: www.earth-syst-dynam.net/2/1/2011/esd-2-1-2011.pdf, accessed 20 May 2015.

Millstein, D. and Menon, S. (2011), 'Regional Climate Consequences of Large-Scale Cool Roof and Photovoltaic Array Deployment', Environmental Research Letters, Vol. 6. DOI:10.1088/1748–9326/6/3/034001.

Mitchell, D. (2008) 'A Note on Rising Food Prices', Policy Research Working Paper 4682, World Bank, Development Prospects Group, online resources: www-wds.worldbank.org/external/default/WDSContentServer/WDSP/IB/2008/07/28/000020439_20080728103002/Rendered/PDF/WP4682.pdf, accessed 15 April 2009.

Mortimer, N. (1991) 'Energy Analysis of Renewable Energy Sources', *Energy Policy* Vol. 19 No. 4: 374–85.

Murphy, D.J. and Hall, C.A.S. (2010). 'Year in Review EROI or Energy Return on (Energy) Invested'. *Annals of the New York Academy of Sciences*, Vol. 1185: 102–18.

Newman, G. (2000) 'Cleaning Up Hydro's Act', *International Water Power & Dam Construction*, August: 14–15.

OECD (1989) 'Environmental Impacts of Renewable Energy', Compass Project Report, Organisation for Economic Co-operation and Development OECD.

Oud, E. (2002) 'The Evolving Context for Hydropower Development.' *Energy Policy* Vol. 30: 1215–23.

Petts, G.E. (1984) *Ecology of Impounded Rivers: Perspectives for Ecological Management*, John Wiley & Sons, Chichester.

Pickard, W.F. (2014), 'Energy Return on Energy Invested (EROEI): A Quintessential but Possibly Inadequate Metric in a Solar Powered World', *Proceedings of the IEEE*, Vol. 102 No. 8.

Pimentel, D., Fast, S., Gallahan, D. and Moran, M. (1983) 'Environmental Risks of Utilizing Crop and Forest Residues for Biomass Energy', *International Energy Conversion Engineering Conference*, Vol. 2: 580–5.

Ranganathan, V., 2000, 'Hydropower and the Environment in India', *Energy Policy* November.

RFA (2008) 'The Gallagher Review of the Indirect Effects of Biofuels Production', Renewable Fuels Agency, online resources: www.unido.org/fileadmin/user_media/UNIDO_Header_Site/Subsites/Green_Industry_Asia_Conference__Maanila_/GC13/Gallagher_Report.pdf, accessed 16 September 2015.

Richards, B.S. and Watt, M.E. (2007) 'Permanently Dispelling a Myth of Photovoltaics via the Adoption of a New Energy Indicator', *Renewable and Sustainable Energy Reviews*, Vol. 11: 162–72.

Richey, J., Melack, J.M., Aufdenkampe, A.K., Ballester, V.M. and Hess, L.L. (2002) 'Outgassing from Amazonian Rivers and Wetlands as a Large Tropical Source of Atmospheric CO_2', *Nature*, Vol. 416: 617–20.

Royal Society of Chemistry (2005) 'Environment, Health and Safety Committee Note on Life Cycle Assessment.' Version 2 RSC, online resources: www.rsc.org/images/LCA%20Revised%20Note%20Final%20140205_tcm18-97943.pdf, accessed 13 January 2009.

Smil, V. (2015) *Power Density: A Key to Understanding Energy Sources and Uses*, Cambridge, MA: MIT Press.

Sta Maria, M. and Jacobson, M. (2009) 'Investigating the Effect of Large Wind Farms on Energy in the Atmosphere', *Energies*, Vol. 2: 816–38.

Stephenson, A. and Mackay, D. (2014) 'Life Cycle Emissions of Biomass Electricity in 2020, Scenarios for Assessing the Greenhouse Gas Impacts and Energy Requirements of Using N. American Woody Biomass for Electricity Generation in the UK,' Department of Energy and Climate Change, July.

Tan, K.T., Lee, K.T. Mohamed, A.R. and Batia, S. (2009) 'Palm Oil: Addressing Issues and Towards Sustainable Development', *Renewable & Sustainable Energy Reviews*, Vol. 13: 420–427.

Trussart, S., Messier, D., Roquest, V. and Akil, S. (2002) 'Hydropower Projects: A Review of Most Effective Mitigation Measures', *Energy Policy*, Vol. 30: 1251–59.

Walker, G. (1995a) 'Energy, Land Use and Renewables: A Changing Agenda', *Land Use Planning*, Vol. 12 No. 1: 3–6.

Walker, G. (1995b) 'Renewable Energy and the Public', *Land Use Policy*, Vol. 2: 49–59.

Wang, C. and Prinn, R.G. (2010) 'Potential Climatic Impacts and Reliability of Very Large-Scale Wind Farms', *Atmospheric Chemistry and Physics*, Vol. 10: 2053–61.

Warren, C.R., Lumsden, C., O'Dowd, S. and Birnie, R.V. (2005) 'Green on Green: Public Perceptions of Wind Power in Scotland and Ireland', *Journal of Environmental Planning and Management*, Vol. 48: 853–75.

Watt Committee (1990) *Renewable Energy Sources*, Elsevier, London and Paris.

WCD (2000) 'Does Hydropower Reduce Greenhouse Gas Emissions?', World Commission on Dams Press Release, 27 November 2000, online resources: www.dams.org/news_events/press357.htm, last accessed 20 July 2009.

WCD 2000b,'Dams and Development. A new framework for decision-making.', World Commission on Dams, Earthscan, ISBN: 1-85383-798-9 paperback, 1-85383-797-0 hardback, London and Sterling, VA.

WEC (2004) *Comparison of Energy Systems Using Life Cycle Assessment*, World Energy Council, London.

Windpower Monthly (1998) 'Windfarm Design at the Touch of a Button', *Windpower Monthly*, September.

WWF (2003) 'Better Management Practices: The Way Forward to a Sustainable Future for the Oil Palm Industry', WWF Malaysia, online resources: http://assets.panda.org/downloads/bmpfinal.pdf, last accessed 2 May 2016.

WWF (2005) 'Agricultural and environment: Palm oil; habitat conversion'. Available at: http://www.panda.org/about_wwf/what_we_do/policy/agriculture_environment/commodities/palm_oil/environmental_impacts/habitat_conversion/index.cfm, last accessed 2 May 2016.

3 Theory

3.1 Introduction

The previous chapter reviewed the literature on environmental impacts of renewable energy, concluding that few systematic physical approaches exist that can compare and help predict impacts for all of the renewable sources. This chapter discusses a theory proposed for such an approach.

3.1.1 How to define impact?

The definition of impact used here is one of changes, and in particular step changes, the pushing of a stable system beyond its normal dynamic into a new state. These changes can be expressed numerically in terms of the extent, the frequency and the rate of change.

But which changes are significant? The state of the natural environment can be said to be maintained by energy flows as described in Table 1.1, and changes to these flows can be expected to result in environmental changes. Natural energy flows can be characterised in terms of their intensity, that is, their power flux density. This could be expected to relate to the amount of work that could be performed per unit area (perpendicular to the flow), and thus to the functions being performed in the natural environment, whether it is the physical processes of geomorphic erosion and deposition, or bio-chemical processes such as photosynthesis.

If the definition of impact used here is change to a given flow of energy, one of the significant changes is likely to be the interception and extraction of energy from the flow. The proportion of the flow that is intercepted and abstracted is likely to have a bearing on the changes to the flow.

Figure 3.1 illustrates this conceptually for a natural energy flow, e.g. a river. The diagram shows energy losses from natural energy flow reducing energy available for functions.

Interception of the flow is likely to cause changes or perturbation, while extraction of energy will leave less available for the functions described in Chapter 1. These functions, such as erosion, transport of sediment and nutrients, maintain the river's environmental conditions. For example, the

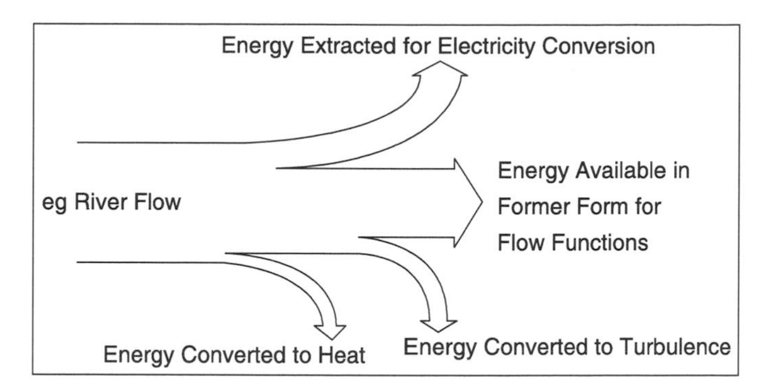

Figure 3.1 Sankey diagram showing energy losses from natural energy flows.

Source: adapted from Clarke (1993).

maintenance of deltas at the mouths of major rivers depends on the supply of silt by the river, and loss of that supply can lead to the erosion of the delta by the action of the sea.

In terms of changes to the flow and the degree of perturbation of the flow, the energy conversion device efficiency is likely to have an influence. Device efficiency is also likely to influence the proportion of the flow that is intercepted and extracted.

An inefficient energy conversion device is likely to produce unwanted by-products, whether it is perturbation of a water flow or chemical by-products in the atmosphere or heat.

The mode of energy conversion itself is likely to relate to the overall impact. Since no energy conversion process, except ultimately that to low-grade heat, can be 100 per cent efficient, and unwanted by-products tend to be produced by left-over perturbed energy flows, the number of different energy conversion stages is likely to be related to the impact experienced, since each step is likely to involve a change to the flow and may generate impacts, cumulatively or independently. It is suggested that the more direct the energy conversion process, the lower the impact will tend to be, all other factors being equal.

3.1.2 Parameters

This interpretation leads to the suggestion that a limited number of parameters could be used to measure these factors and thus provide an assessment of the likelihood of impacts.

1 Power flux density is the first parameter identified here, which might indicate the degree of activity or work performed on or in a certain area. The extent or size of the natural energy flow will indicate the total extent

of work done, which might be related to the cross-sectional area of a river, or air flow, or alternatively of solar radiation flows. However, it is the intensity of the work performed per unit of area, or the force × distance applied per square metre, that can indicate the type and degree of work performed.

2. The proportion of the flow intercepted and then abstracted should be a significant parameter, in terms of the effect on the functions performed by the energy flow in the natural world.
3. A third parameter to consider is the converter efficiency, since this may have a bearing on the degree of perturbation of a natural energy flow, in terms of unwanted by-products.
4. The number of conversions involved in the energy conversion process could prove to be a significant indicator of impact.

In this book four variables are labelled:

d Power flux density
p Proportion of the flow intercepted and then abstracted
e Efficiency of the converter device
n Number of energy conversions

3.2 Energy/power flux density

A fundamental question with far-reaching implications is 'How intense or diffuse is the renewable energy source flow?' Why should this be a fundamental question? Because if it has been established that renewable energy flows power and maintain the existing environment, i.e. they perform 'work' to do this, a key metric should be how much work and spread over what area. If renewable energy flows maintain the existing environment, then impact can be defined as changes to those flows, resulting in either a deficiency of work performed, or in 'unwanted work', i.e. unwanted by-products. It is the *intensity* of that work which is significant, as the examples below, such as solar radiation compared to lasers, show. As power flux density increases, so the effect is to cause a change in the nature of the work, i.e. the energy conversion performed. Of course, the form of the energy flow, whether electromagnetic radiation, mechanical or heat flows, is also significant.

3.2.1 Terminology

Some confusion or debate may arise concerning the terms employed here and the units of measurement used. The term *energy* flux might be used because energy is usually defined as 'the capacity to do work'. The flows of energy described here represent the capacity to do work. Since what is being described here is a flow of energy, the unit of measurement would be joules per second; but there is already a unit for that, i.e. the watt. Watts are used to describe *the rate at which work is done*, that is a measure of power. A flux or flow of necessity involves the dimension of time and this is therefore the *rate* of flow. However, doing work implies a conversion of energy from one form into another. But that is not the issue here; the 'work' or transformation of the energy form has either already occurred and/or is about to occur again. The result is that the energy flows are measured here in watts per square metre, which purists may argue is the wrong term, or alternatively requires the term 'power flux density'. In the literature this term, or just 'power density', is often used; in practice and in this book watts are used somewhat interchangeably with joules per second. However, for the sake of clarity, the term 'power flux density' is used here.

3.2.2 Power flux density

The concept of power flux density is defined here as the rate of energy flow per square metre across a hypothetical plane across the energy flow perpendicular to it, as shown in Figure 3.2.

The eight primary renewable energy sources identified earlier vary considerably in how diffuse or intense they are. All of these energy flows are ultimately derived from the sun, apart from tidal and geothermal flows, which are derived respectively from the gravitational pull of the moon (and sun) and the radioactive decay of rocks, as well as residual heat from the formation of the earth (Alexander in Boyle 1996).

It is evident, then, that solar energy arriving at the earth's surface at a maximum of ~1 kW m^{-2} is then concentrated through successive conversions into flows that are generally more intense (though smaller in size of total resource), starting with wind derived from heat from solar radiation, and pressure differences across the world, going on to ocean waves, derived from wind, and further to wind and thermally driven ocean currents. Solar radiation heating the atmosphere and the earth's surface evaporates water in

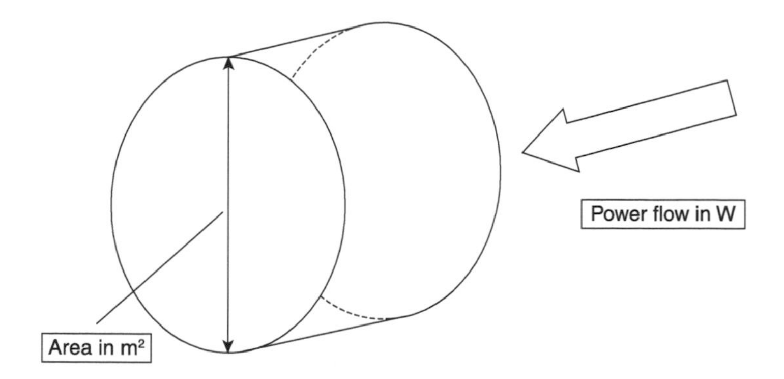

Figure 3.2 Diagram showing components of power flux density of natural energy flow.

the hydrological cycle, which through falling water can provide hydro electric power (HEP).

Biomass energy is somewhat of an exception here since, on the one hand, at the conversion stage in a photosynthesising plant, it has a very low total power flux density of 450–150 GJ ha^{-1} p.a. or 1.43–0.48 W m^{-2}, being in the UK only about 0.5 per cent efficient in conversion and storage of solar energy (Ramage and Scurlock 1996: 145), and on the other hand, the stored carbohydrates themselves represent flows of energy in the soil with a potentially higher power flux density.

Ocean currents, mentioned above, can be wind-derived, e.g. the Gulf Stream, but also thermo-haline in origin, produced by temperature and density variations due to salt concentration differences (Brandon and Smith 2003: 215). An example here is the North Atlantic Conveyor current. Closer to land, the tides are the main cause of currents, and these currents can become amplified by the shape of coastlines and estuaries as they are 'squeezed' into narrow channels, or resonate in funnel-shaped estuaries (Charlier 2003). There is apparently no clear boundary between these types of currents, which connect together, though each can be identified.

Geothermal energy has very low average background power flux density of 0.06 W m^{-2} (Boyle 1996), but can have a high power flux density at naturally occurring vents such as geysers.

Since the power flux density of the different sources varies considerably, by up to a factor of at least ~1,750, this has profound implications for the area required and potentially for land use. See Figure 3.3 and Table 3.1 for the power flux densities of eight different primary renewable energy flows. The intensity and variability of different natural energy flows is distributed unevenly geographically and in time, as well as between the different sources. Some examples of this are given in the next section.

Intensity of natural energy flows

The intensity of natural energy flows varies in nature both spatially and temporally. Natural variability in the intensity of such energy flows (i.e. the power flux density), can be compared with other examples of concentrated energy flows for illustration.

For example, whereas normal sunlight intensity has a maximum of ~1 kW m^{-2} at the earth's surface and 1.365 kW m^{-2} at the top of the earth's atmosphere (Boyle 1996: 96), light can be concentrated in laser light beams with a power flux density of beyond, for example, 10^6 kW m^{-2} (Key *et al.* 1972). By concentrating the energy of the beam in a very small area, the work performed can include cutting, vaporising and welding. Visible light wavelengths vary from ~0.39 μm to ~0.7 μm, with peak energy transmitted at ~0.45 μm to ~5 μm (Boyle 1996: 44). Ultraviolet wavelengths of light with a shorter wavelength, e.g. 0.35 μm, have more energy than visible frequencies. The atmosphere and ozone layer shield the surface from higher ultraviolet fluxes, though more of

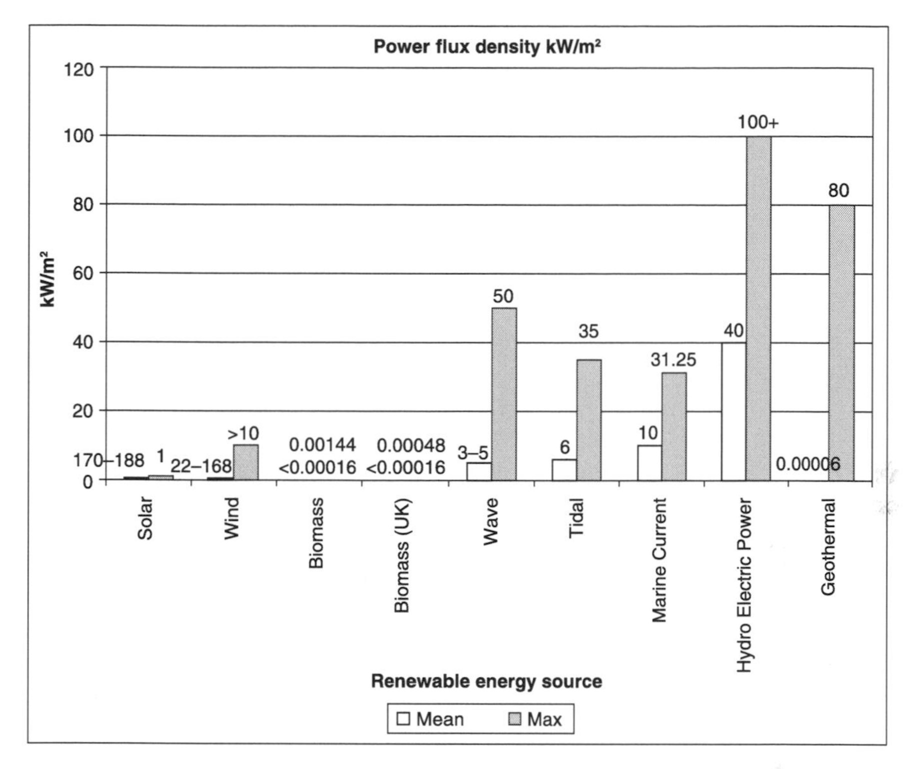

Figure 3.3 Renewable power flux density before concentration at the converter.

Source: adapted from Clarke (1993).

Notes: solar: 1 kW m^{-2} (Twidell and Weir 1986); marine current 10 MW km^{-2} (Fraenkel 2007). Figures are indicative as there may be wide individual variations. Solar: 3–30 MJ m^{-2} per day mean, <1kW m^{-2} max. Wind: 3.28ms^{-1} (land), 6.5 ms^{-1} (sea), mean, and 25 ms^{-1} max. Biomass: assumed 320 W solar radiation max absorption, <0.5 per cent efficiency, remainder of energy used in evapo-transpiration and heat. 16 W m^{-2} leaf surface mean energy flow in sunlight. Wave: 20–40 kW m^{-1} mean, N. Atlantic, assuming wave depth is 10 m, max. wave height, e.g. once per 50 years, may be up to 50 times height, but depth of <100 m. Energy density also dependent on sea depth. HEP assumes medium-high head site. Geothermal: 0.06 W m^{-2} mean earth surface value, max. e.g. geysers.

these are experienced at higher altitudes. As the angle of the sun and cloud cover change, the energy flux changes, decreasing to near zero at night.

Wind energy also varies considerably, from about 40 W m^{-2} on average, based on 4 m s^{-1} average wind speed (at the converter) across the earth (WEC 2004), to locations with 615 W m^{-2} at an average speed of 10 m s^{-1}, and up to 56 kW m^{-2} at 45 m s^{-1} (based on power P (in watts) = ½ × 1.23 × A × V^3) or greater in storms or hurricanes. At such power levels, considerable damage can be caused.

Similarly, rainfall varies both spatially and temporally between about 0 mm p.a. and 10,000 mm p.a. (Brandon and Smith 2003); many rivers have cyclic

Table 3.1 Power flux density for renewable energy sources, at the converter

	Mean power flux density	*Maximum power flux density*
Solar	100W–~300 W m^{-2}	1000 W m^{-2}
Biomass (UK)	0.16 W m^{-2}	0.57 W m^{-2}
Wind	168 W m^{-2}	10–12 kW m^{-2}
Hydro	40 kW m^{-2}	100+ or >100 kW m^{-2}
Wave	3–5 kW m^{-2}	50+ or >50 kW m^{-2}
Tidal	6 kW m^{-2}	30+ or >30 kW m^{-2}
Marine current	10 kW m^{-2}	31.25 kW m^{-2}
Geothermal	0.06 W m^{-2}	80 kWm^{-2}

Note: figures are indicative only and cannot represent the wide ranges at individual sites.

flood flow periods, when flows can be several magnitudes greater than continuous flows. As the gradient of rivers is not constant, the intensity of the energy flow will depend on the gradient at any location along the river. The example of a waterfall can be considered compared to a river in its flood plain with, e.g. high head hydro electric schemes compared to low head hydro power and resulting differences in intensity of energy flows. High-intensity flows can transport larger-size sediments, even pebbles and boulders, and erode river-bed rocks.

3.3 Land use

One of the most evident relationships between power flux density and environmental impact may be the issue of the area required by any renewable energy source, per unit of energy produced, or on a larger scale, the use of land or space required. 'Land use' is the term usually applied to the type of use of land by humans. Here it is applied to the amount of land used by each source and technology per kW of power.

The land used and some of the factors and issues involved are discussed below for these sources:

- solar
- biomass
- wind
- hydro
- wave
- tidal
- marine current
- geothermal

For solar, the power flux density, or intensity, can be fairly easily related to the area of collector required, and also to general notions of land use. Since power flux density is here defined as the watts per square metre (W m^{-2}) of an imaginary collector perpendicular to the ambient energy flow, it is apparent how the density is related to collector area; a greater power flux density will require less collector area per energy unit. If the flow is horizontal to the earth's surface, then the collector angle will be perpendicular, i.e. vertical, and the area it occupies will be in a vertical plane. Hence land use will tend to be less. Where the renewable energy flow's angle relative to the earth's surface is greater than zero degrees, i.e. tilted, the collector angle will occupy space in a horizontal plane in addition to a vertical one.

Solar and geothermal flows are generally from a vertical (though varying in the case of solar) direction, or perpendicular to the earth's surface, while water flows and air flows are generally more horizontal or nearly parallel to the earth's surface. So it would follow that direct solar, biomass and geothermal collectors tend to occupy horizontal areas, i.e. using land area. Wind, hydro, wave and tidal collectors tend to occupy space in a vertical plane, and relatively speaking may not occupy very much land area. Of course, the associated ancillary conversion and energy storage facilities such as reservoirs may well occupy considerable land area which needs to be taken into account too.

3.3.1 Land use and power flux density

Do typically quoted land use values for each renewable source tend to bear out the relationship with the power flux density? Those cited by Gagnon *et al.* (2002), shown in Figure 3.4, only do so to an extent, in that biomass plantations require the largest land areas and have the lowest power flux density, and the hydro run of river requires the lowest land area, and has a relatively high power flux density, as might be expected. The figures for hydro with reservoir range from about 5 km^2 to 200 km^2/TWh, a surprisingly large figure considering that wind power is allocated a ~30–130 km^2/TWh range, and again surprisingly and questionably, this is higher than the figure for solar PV at ~30–45 km^2/TWh.

Some of these results would appear to contradict the hypothesis here, being inverted in relation to power flux density at the converter and may be averaged over total land areas. Unfortunately due to the paucity of references and lack of explanation of methods used to generate these data, only suggestions can be made for the difference in results. The authors do, however, caution against simple interpretation of the data as 'it does not consider the intensity of impact nor the degree of compatibility of generation options with other land use'.

Moreover, the authors point out that only the direct use of land is considered and that accounting for indirect land 'use', e.g. from acid deposition from coal burning, or the longevity of nuclear waste storage periods can alter the land use figures by one or even two orders of magnitude.

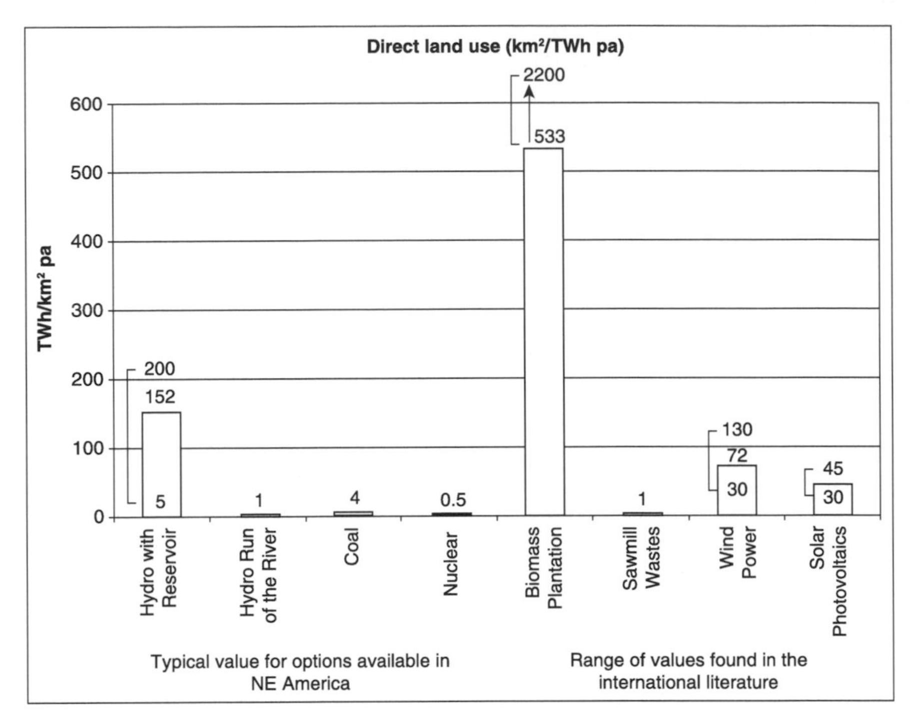

Figure 3.4 Land use of renewable and conventional energy sources data from international literature.

Source: adapted from Gagnon *et al.* (2002: figure 4).

Notes: range of values found in the international literature: (1) hydro with reservoir for base and peakload options: ~5–200 km²/TWh; (2) hydro run of river: 1 km²/TWh; (3) coal: 4 km²/TWh; (4) nuclear: 0.5 km²/TWh; (5) biomass plantation: 533–2,200 km²/TWh; (6) sawmill wastes: 1 km²/TWh. Intermittent options that need backup generation: wind power: ~30–~130 km²/TWh; solar photovoltaic ~30–45 km²/TWh.

Hydropower's results vary significantly due to site-specific conditions, they state, as well as the multi-purpose nature of many schemes, together with the storage function of the reservoir. However, as few references are given, the data sources and method of calculation are not apparent.

The high figures for wind energy at 72 km² per TWh p.a., or ~2.3 $m^2 W^{-1}$ (0.44W m^{-2}) may be due to using low wind speed examples, and land area across which turbines are spread, rather than actual land occupied. The authors state that

> for wind power the land around the windmills can still be used for agriculture . . . that solar energy can be developed on rooftops or arid areas without agriculture . . . and that hydropower can be developed in mountainous or cold climate areas.

It would appear that a rather imprecise interpretation of 'land used' may account for the figures provided. Possibly no distinction has been made between land actually occupied and land affected by a renewable collector device or scheme.

A more recent study by Ong *et al.* (2013), a report by the US NREL on land use requirements for solar power plants, cites considerably lower land use figures for a variety of technologies than Gagnon *et al.* above. Small installations of solar PV of 1–5 MW size are cited at an average of 4.1 acres/GWh p.a. or 16.59 km^2/TWh p.a. Large solar PV of >20 MW occupy an average of 3.4 acres/GWh per year, or 13.76 km^2/TWh p.a., and solar CSP (parabolic trough, power tower and Fresnel types) an average of 3.5 acres/GWh p.a., or 14.16 km^2/TWh p.a. Solar power plant total land use ranges from 3 acres to 5.5 acres/GWh per year for small, two-axis, flat PV power plants, with an average of 3.5 acres/GWh per year, according to this study.

3.3.2 Diffuseness

MacKay, in his estimation of UK renewable energy resources, cites the power per unit land or water area, and argues that 'renewable facilities have to be country sized because all renewable sources are so diffuse' (MacKay 2008b). MacKay (2008b) cites land use in watts per square metre for different renewable energy sources, as shown in Table 3.2.

Here, MacKay is citing the estimated land area for each source, and averaging the power density over the estimation of practically available land from the UK's area of 245,670 km^2. This method differs from that of this book, which considers the power flux density per square metre of that part of the flow which is harnessed, and the power density per square metre at the converter itself.

MacKay's concern is to estimate the overall resources available for the UK as a whole, and its population, as opposed to comparing the power flux density at the converter, with the resulting environmental impacts.

Table 3.2 Average overall power per unit of area of total resource

Wind	2 W m^{-2}
Offshore wind	3 W m^{-2}
Tidal pools	3 W m^{-2}
Tidal stream	6 W m^{-2}
Solar PV panels	5–20 W m^{-2}
Plants	0.5 W m^{-2}
Rain-water (highlands)	0.24 W m^{-2}
Hydroelectric facility	11 W m^{-2}
Solar chimney	0.1 W m^{-2}
Ocean thermal	5 W m^{-2}
Concentrating solar power (desert)	15 W m^{-2}

Source: adapted from MacKay (2008b: table 3)

These two objectives are related in that the practical resource size is determined by constraints resulting from environmental impacts. MacKay does not investigate the link between power flux density at the converter and environmental impacts and constraints. The assumption is that this is relatively fixed; his objective appears to be to produce viable plans at a certain moment in time, hence the 'state of development' is taken as read. However, the estimations of power per unit area are relevant to this study in terms of the maximum packing density of energy converters over a larger area. This topic is discussed for wind energy in Chapter 9.

3.3.3 Land use for HEP schemes

Land use per TWh might be expected to be inversely related to power flux density, in that the higher the power flux density, the lower the area required. However, land use for HEP schemes comprises mainly the reservoir area, which is a function of the head height chosen and the river gradient, which will determine how long the reservoir is, as well as the topography of the valley, together with breadth and steepness of the sides. Together with flow rate, these factors then determine the land use per unit of energy generated. It is mainly the need for large storage capacity to even out the high seasonal flow variation, as well as achieving a head height that is as large as possible, that results in large land area use per unit of energy production.

A formula for land use for hydro electric schemes might be proposed:

$$A/\text{unit of energy p.a. (e.g. TWh p.a.)} = B \times (H/H\,(s \times g \times Q \times \rho \times \eta))$$

Where:
A = land area
B = reservoir breadth
H = head height
s = slope or river gradient
g = acceleration due to gravity
Q = flow rate
ρ = density
η = coefficient of performance (overall conversion efficiency).

Dimensional analysis

- Area is expressed as m × m or $[L]^2$.
- Unit of energy produced per unit of time is expressed as power or $[M] \times [L]^2 \times [T]^{-3}$.
- Breadth is expressed in m or [L].
- Head is expressed in m or [L].
- Slope is expressed as m/m or $[L]\,[L]^{-1}$.
- Acceleration due to gravity is expressed as m s^{-2} or $[L]\,[T]^{-2}$.

- Flow rate is expressed as $m^3\ s^{-1}$ or $[L]^3\ [T]^{-1}$.
- Density is expressed as ρ = kg m^{-3} or $[M]\ [L]^{-3}$
- η = dimensionless

So:

$$[L]^2/[M] \times [L]^2 \times [T]^{-3} = [L]\ [L]/[L]\ [L]\ [L]^{-1}\ [L]\ [T]^{-2}\ [L]^3\ [T]^{-1}\ [M]\ [L]^{-3}$$

This shows that the equation balances in terms of dimensions.

It might be expected that land use per unit of energy will be lower for schemes with greater river gradients, higher mean flow rates, as well as narrower valleys and higher coefficients of performance.

3.3.4 Space occupation

As stated, energy converters take up space both in the horizontal plane and also in the vertical plane. However, the different technologies and sources require varying volumes. For instance, the very thin converters of solar PV technology can be contrasted with wind turbines that occupy a spherical volume (in order for the blades to be oriented or yawed through 360° some metres above the ground. Other technologies with multiple-stage conversion equally require a volume of space. As a consequence, solar PV panels can be substituted for roof tiles or slates, with very little effective impact despite the low power flux density of solar radiation (ETSU 1994).

3.4 Proportion

There are different ways of employing the concept of proportion of energy flow extracted; for example, at different scales. The concept could apply to a technology, e.g. wind energy, or it could apply to the design for a single development, e.g. an HEP scheme, or to a whole river. Alternatively, the concept of proportion can refer to the fraction of a total resource extracted or intercepted in a region, or a nation, or even a whole continent.

As stated, if it is accepted that natural energy flows power the natural environment, then extracting energy from those flows might be assumed to affect processes in the natural environment. A reasonable expectation could be that the proportion of energy extracted will be related to the impacts.

As an illustration of impacts from intercepting energy flows, one might consider the impacts from an extreme situation, interception of the total flow. For instance, the effect of 100 per cent interception of light for solar PV could be considered, so that no light reached the earth's surface. Some examples of the concept of interception of *proportions* of the energy flow are given here, illustrated through some extreme polarities.

Solar: solar radiation flows arrive at the surface of the earth, providing light in the visible waveband. This radiation flow is moderated by the ozone layer, the atmosphere, cloud cover, dust and finally by vegetation (e.g. 1.365 kW m^{-2} maximum outside the earth's atmosphere and up to 1 kW m^{-2} maximum on the surface) (Boyle 1996: 96). Solar collectors intercept a proportion of this flow. If the total solar flow interception were to be 100 per cent (i.e. a 100 per cent efficient black body absorber) covering 100 per cent of the earth's surface, the result would be darkness at the surface of the earth, thus preventing photosynthesis, the primary basis for life, from occurring. One hundred per cent earth surface coverage would result in permanent darkness, versus normal conditions. A 100 per cent interception rate can be compared with the natural moderation by the atmosphere, i.e. the spectral power distribution, as a result of air mass distribution (Boyle 1996: 96), cloud cover and dust absorption. Even if a lesser area were to be covered, sufficient light could be intercepted to prevent plants growing in an area. At still lower percentages of coverage, the result would merely be increased shade, and a decrease in light intensity with resulting decrease in heat.

Biomass: biomatter carbohydrate flows through the biosphere via decomposition and soil, but can be intercepted by harvesting for combustion. If *all* the biomatter were harvested, including the roots, and combusted so that no organic carbon reached the soil, the result would deprive soils of organic nutrients, carbon, hydrogen, and nitrogen particularly. A process akin to this can occur when deforestation occurs in hot, dry, tropical areas, leading to desertification.

Wind: wind turbines function through slowing down the wind slightly. If the air flows of weather systems could be entirely halted, there would be considerable impact on the weather system, rendering the air flow and circulation component inoperative.

Water Flows: flowing water under the influence of gravity is exploited by HEP systems, which abstract some of the kinetic energy by slowing down the water speed. If the water speed could be entirely halted, very considerable changes to the flow and water system would be experienced. For an example, the freezing of an entire river system, as occurred in the ice ages, has major impacts on the ecology of the surrounding ecosystem, and the relevant part of the hydrological system, aside from changes in temperature. However, it is apparent that if the water movement were entirely halted, then that part of the hydrological system would cease to function. More realistically it might be assumed that if a high proportion of the flow energy is extracted, there will be considerable changes in the river system. Since the river is an integral part of ecological systems and has a wider longer-term influence on whole environments, this change can be defined as environmental impact.

The concept of *proportion* might also be employed in relation to the fraction of a total resource developed. For example, Bartle (2002), in a paper on HEP and development, cites Africa as having developed only 4 per cent of the technical resource.

3.5 Efficiency

The efficiency of energy conversion may be defined as the electrical, fuel or heat energy output from the converter divided by the theoretical energy available to the converter, in a given flow. This can be significant to environmental impact in several ways, affecting the proportion of energy that can be extracted, the possible perturbation of the flow and in some cases the formation of by-products. High-efficiency conversion capability enables a large proportion of a flow to be abstracted, an example of which is the conversion efficiency of high- and medium-head hydro electric plant of up to ~90 per cent. By contrast, wind turbines and marine current turbines are subject to a theoretical maximum efficiency limit of ~59 per cent, known as the Betz limit (Twidell and Weir 1986; Betz 1920), since such turbines are sited in an extended fluid stream. Slowing down the flow further would be counterproductive, since the flow at the converter would be eventually halted and would bypass the obstruction created. In practice, actual conversion efficiencies will be lower still than the 59 per cent of the Betz limit. In addition, wind and marine current turbines need to be spaced out from each other to avoid being in each other's wind or current 'shadow'. As a consequence, the overall proportion of the energy flow that can be intercepted and abstracted is much lower than that of hydro electric schemes.

Similarly, the efficiency of wave energy converters could be related to the energy left in the wave once it had passed a converter, and thus the size of the wave that for example reaches the shore. Different operational principles are employed in the different wave energy converters, with resulting varying efficiencies. Figure 3.5 shows indicative renewable energy converter efficiencies for the main sources.

3.6 Number of conversions

The number of conversions describes how many times the energy flow is transformed into other forms, for example from kinetic energy to potential energy, or from thermal energy to rotational kinetic energy. This variable appears to be related to the form of the energy flow rather than the power flux density. One reason that transformations of energy form, such as from kinetic to potential, are likely to be significant in environmental impact is that they represent discontinuities in the energy flow functions. For instance, the transport of sediment by the kinetic energy of flowing water will be interrupted when that kinetic form is converted into potential form in a reservoir. The water speed will drop and the sediment in suspension will be deposited.

Such conversions of energy form, and ensuing interruptions to functions, are not just limited to HEP, but are likely to occur in some of the other renewable energy source conversion technologies. For example, concentrating solar thermal power stations convert the radiant solar beams to heat form and then to stored energy potential in raising steam.

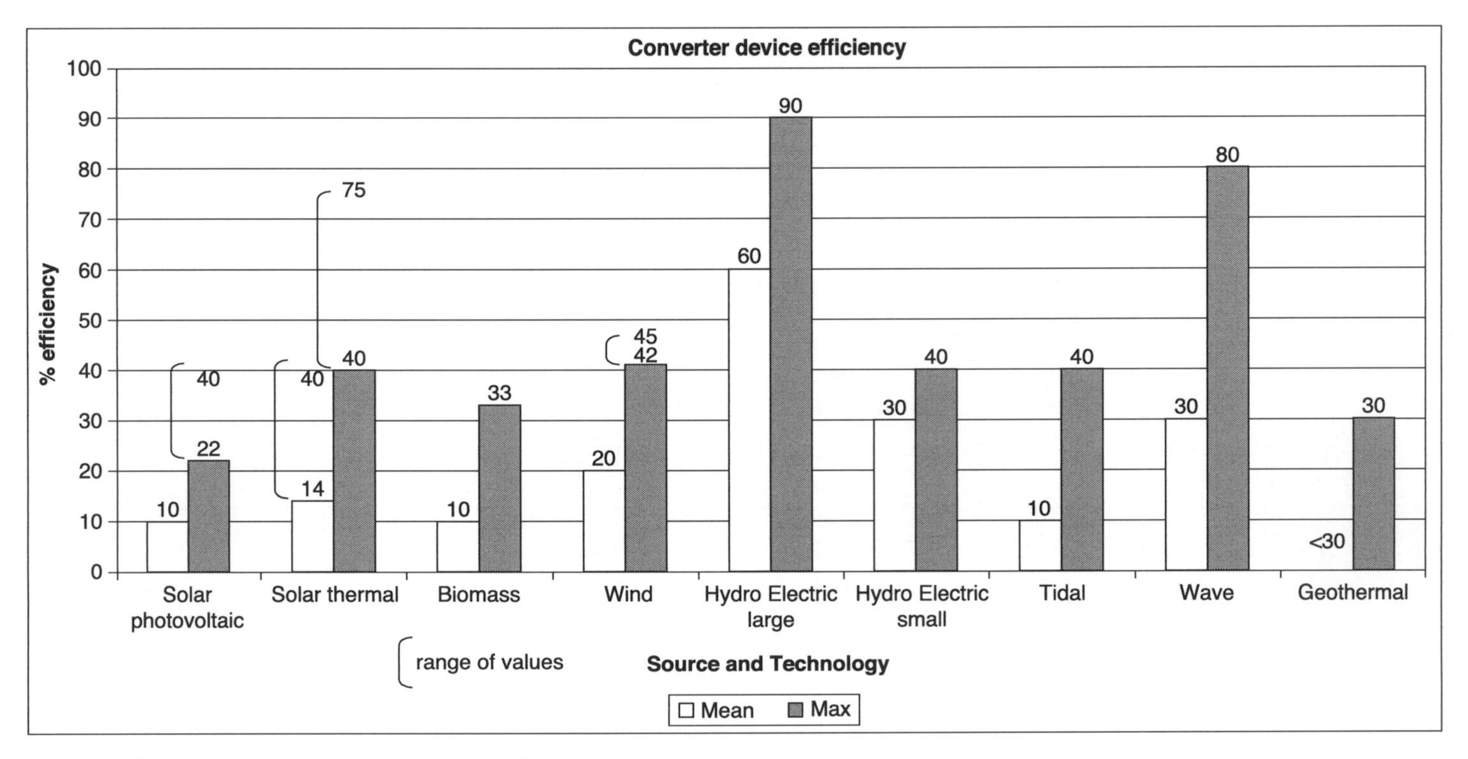

Figure 3.5 Renewable energy converter device efficiency.

Source: adapted from Clarke (1993).

Note: definitions of efficiency vary with technology.

At this stage, a distinction should be made between the number of conversions enacted on a natural flow such as a river, before extraction of energy, and those conversions carried out to the energy extracted from a flow which are contained within the 'black box' of the converter device. The number of conversions enacted on the natural energy flow may be more significant, since this part of the flow continues to perform functions in the natural environment.

3.7 High- and low-grade energy sources

Of relevance to the parameters proposed above is the concept of high- and low-grade sources (Alexander 1996: 7–8). 'High-grade' energy sources can be thought of as highly organised, low entropy, lending themselves to efficient conversion into other high-grade forms, such as electricity. 'Low-grade' sources can be conceived of as less organised, high entropy, resulting in low energy conversion efficiency to other higher-grade forms, for example lower temperature heat energy, where the energy is contained in the random oscillatory kinetic motion of molecules. The latter thus requires multiple conversion stages to be transformed to the 'high-grade' form of electricity, with resulting loss of overall efficiency. This is governed by the Second Law of Thermodynamics (Clausius 1856). However, when comparing conversion efficiencies of energy flows of forms as different as solar radiation, flowing water or thermal energy, this raises issues as to the extent to which thermodynamics can be applied at the quantum level, which is beyond the scope of this book.

The term 'high grade' can also, in the case of heat sources such as in geothermal energy, refer to the power flux density for flows of energy – the higher the power flux density the higher the grade. This usage of the term relates essentially to thermal energy conversion, where increases in temperature difference ΔT increase the efficiency of conversion.

For static stores of energy such as fuels or steam, 'high-grade' forms refer to the energy density or energy per cubic metre or per unit of mass, i.e. $J\ m^{-3}$ or $J\ t^{-1}$.

The importance of the concept lies in the conversion efficiencies possible resulting from the grade of energy. Conversion from one high-grade source to another high-grade form is often possible at high efficiency, for example falling water (HEP) to electricity at <90 per cent. Conversion from a high-grade source to a lower grade is also possible at high efficiency, but conversion from a low-grade source to a higher one will only be possible at low efficiency. Thus thermal conversions to electricity rarely have efficiencies greater than 50 per cent, usually ~30 per cent, or less depending on the temperature difference. To achieve a theoretical 80 per cent conversion efficiency would require temperature differences of well over 1,000 °C, which is unlikely in most cases, as yet, to be a practical option.

As stated above, if lower efficiencies are likely to result in more unwanted by-products, i.e. the likelihood of environmental changes and hence impacts, then limitations of conversion efficiency are key to explaining impacts.

3.8 Interim conclusions

The hypothesis proposed in this book holds that the power flux density of a natural energy flow is a significant variable in determining the environmental impact from renewable energy sources. This is due to the following:

1. The intensity of the functions performed in nature by the energy flow; the higher the power flux density the higher the rate of work of the functions performed per unit area. The power flux density (rate of work performed) can determine the type of change effected, whether mechanical, chemical or even at an atomic level for nuclear reactions.
2. The relationship between power flux density and land use – the higher the power flux density, the lower the land use; fewer collectors are required per unit of energy, resulting in less usage of land area.

While power flux density is proposed as a fundamental factor in characterising impact from renewable energy sources, it cannot serve to explain all impacts or the degree of impact.

Therefore, in addition to this variable, the proportion (or ratio) of energy extracted from a natural energy flow is proposed as a parameter. The degree of impact caused from extraction of energy from a natural energy flow could, it is suggested, be related to the proportion of energy extracted. Impacts would then result from the changes caused by reduced energy available for natural functions.

Following on from the proportion of energy extracted, another related parameter proposed here is the efficiency of conversion:

3. The relationship between power flux density and efficiency: higher power flux densities can lead to greater conversion efficiencies, e.g. in the case of thermal energy conversion.

The parameter of proportion extracted can be used to explain the limits to the proportion of energy that it is possible to extract, and also help account for aspects of impact due to unwanted by-products.

While these further parameters can help to explain impacts, they still cannot provide a full model of environmental changes caused. Many changes can be effected to natural energy flows without energy actually being extracted from the system. These changes are to the form of energy, e.g. conversion from kinetic to potential, or from radiative to thermal. Such conversions can cause discontinuities to the functions of the natural energy flow, resulting in impacts. The number of such conversions to the form of the energy flow is thus proposed as another parameter.

While four different variables are described here, they are not independent of each other. In addition, natural energy flows normally drive more than one function, e.g. rivers transport sediment, and also perform drainage functions as well as a host of other functions. Both of these functions depend on flowing water, i.e. kinetic energy caused by the influence of gravity. Again, some of these functions are more energy dependent than others. But power flux density, it is argued, might be the most significant variable since the proportion of energy capture from a particular renewable source appears to depend in part on the power density of the flow.

Figure 3.6 shows some proposed typical proportions of flow intercepted/captured compared to power flux density, assembled by the author.

The proportion appears to be highest for the highest power flux density sources, e.g. HEP where the proportion intercepted can be 100 per cent of the flow and the portion captured can be 80–90 per cent, as shown in Figure 3.6. The nature of the well-defined flows that constitute rivers can be contrasted with much broader flows of air currents with less

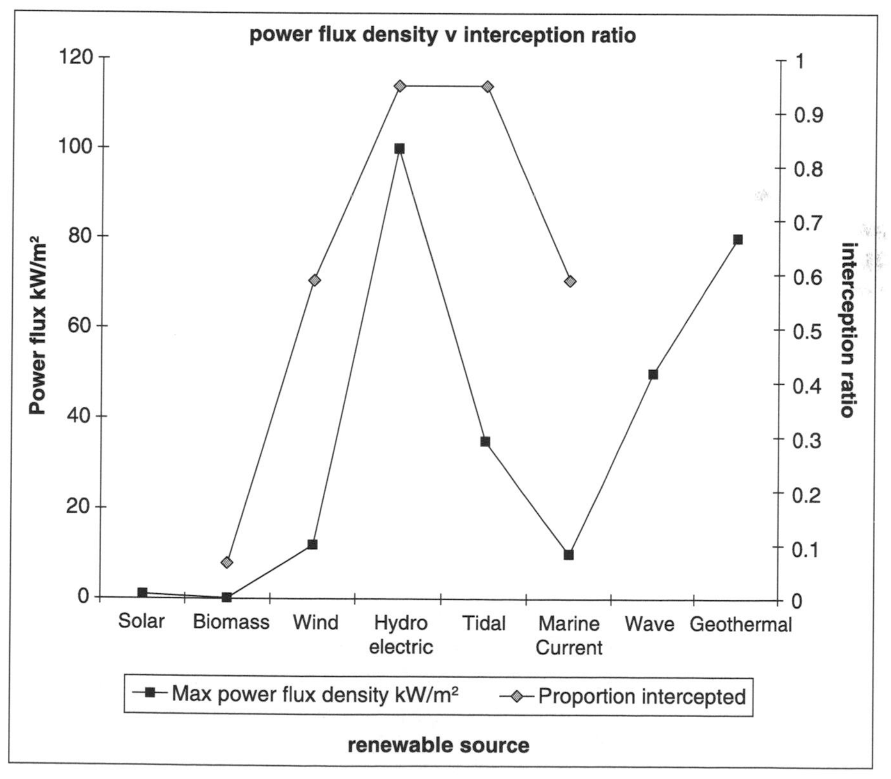

Figure 3.6 General power flux density and interception ratio for renewable energy sources.

Note: figures are indicative as proportion ratios can be at the converter or of total flow.

well-defined boundaries, which will by their nature be less able to be fully intercepted.

For wind energy the maximum theoretical efficiency is expressed by the Betz limit, at 59.3 per cent (Betz 1920, 1966). This is the theoretical limit of energy abstraction in a continuous flow. If a higher proportion of extraction is attempted, then the flow itself is slowed to such an extent that overall still less energy can be extracted. This limit also applies to marine current turbines.

A fifth new parameter can be added here: the changes, i.e. increases/decreases, in the power flux density of the natural energy flow. Such changes could lead to alterations to the flow functions.

The parameters described above – (1) power flux density, (2) proportion extracted, (3) efficiency, (4) number of conversions and (5) changes in the power flux density – may then provide a more complete and satisfactory model of environmental impact from renewable energy sources.

3.8.1 Questions raised by the hypothesis

The hypothesis is that a relationship exists between the environmental impact of renewable energy sources and the power flux density, but that it may not be a simple one in all respects. The hypothesis parameters have been explained above. However, this raises various questions. First, concerning power flux density, does higher power flux density of a natural energy flow really result in a higher rate of work of functions performed? If so what evidence is there for this? Second, concerning a fundamental resource, land use, does higher power flux density for a renewable energy source actually result in a lower land use per unit of energy? Third, concerning the proportion of energy extracted from a renewable energy flow, do impacts result from the reduced energy available for natural functions and is this in proportion to the ratio extracted? Concerning the efficiency of conversion by the converter device, does the efficiency of energy conversion affect the proportion that can be abstracted? Does the efficiency of energy conversion affect the production of unwanted by-products? Concerning the number of energy conversions, are the impacts proportional to the number of conversions to the form of the energy flow? These questions can be further broken down into the different forms of impact, the different renewable energy sources, and the type of relationship with power flux density. Is it true that the greater the power flux density, the greater the number and intensity of the functions performed by the flow? Does a lower power flux density result in fewer functions being performed by the flow? Is the relationship linear or geometric, exponential or some other? What limits might apply to this relationship?

As described in Chapter 2, there is a considerable literature on the effects and impacts of renewable energy sources. Summaries of impacts for the different sources are given in later chapters. The relationship of these impacts to the power flux density might be revealed by ascertaining in some detail the mechanisms whereby the existing environment is powered and maintained by the flows of energy through it. The degree of interception of, or interference with, these energy flows, might reflect the impacts experienced.

However, it would require a major data gathering and analysis effort to try to explore and fully test this relationship in detail for all renewables. This book therefore focuses in detail on one example, HEP, attempting to test whether the hypothesis applies in that case. It then assesses whether this can provide generalisable results, which can hold for other renewables.

3.8.2 Rationale for use of HEP to test the hypothesis

The choice of example was based on the need for extensive existing accessible data, backed by a large body of theoretical knowledge and operational experience.

It was concluded that such characteristics could be represented by HEP. This renewable energy source is the most mature of the newer renewable energy sources and technologies, having been first used in the 1880s (Ramage 1996). As well as having more than 100 years of experience, it is the most widely used of the newer renewable energy technologies, supplying ~2.3 per cent of world primary energy and ~16 per cent of world electricity (IEA 2014). HEP has been described as the only mature renewable energy technology in widespread use (Bartle 2002). HEP developments have been carried out all over the world in a variety of different environments and conditions. HEP has also been deployed on a hugely varying scale, from just several kilowatts in capacity, to the largest power stations in the world at 22.5 GW capacity (Water Power and Dam Construction 2007; Freer 2001), with head heights from ~1–2 m to several hundred to 1,000 metres. Furthermore, much research, modelling and theoretical work has been carried out on the subject of rivers as well as on HEP's environmental effects.

The task could be aided by the fact that river flows have been reliably recorded for over 130 years, while there are flow records dating back thousands of years in the case of the Nile. Extensive flow data exist for a variety of rivers and watersheds worldwide.

3.8.3 Impacts and changes: discussion on the effects of HEP

The impacts of HEP, which have been well documented, need to be measured. In Chapter 1 a definition of 'impacts' as changes was given. Most environmental conditions are in a state of dynamic equilibrium (Petts 1984), and are constantly changing within a certain range, but overall maintaining their relatively steady state. Impacts, by contrast, can be defined as changes which push a system from one equilibrium state to another; that is, they are changes of sufficient magnitude and rate to alter the state of dynamic equilibrium (Catlow and Thirlwall 1976). An example of this for river systems would be the adoption of a new channel course after major floods.

Such changes can be contrasted with the more continuous but damaging unwanted by-products such as chemical pollution, e.g. gaseous emissions from fossil fuel combustion or noise emissions from energy conversion processes. These environmental effects can be considered more as chronic effects, for example chemical pollution such as acid nitric oxide emissions may not be sufficient to cause the cessation of animal or plant life, but imposes continuous damage to respiration and metabolism, reducing health and growth.

Rivers are also responsible for continuous processes such as erosion and transport and deposition of sediment, which over time will affect the morphology of the areas they flow through. Slow but continuous changes such as the building up of deltas are then another natural function of the energy flow; in this case interrupting the flow and halting the processes of geomorphology (i.e. sediment transport here) would constitute the change. So the halting of an existing process of slow, stable change would be defined as an impact.

Applying the question of whether the power flux density is proportional to the functions, for HEP, some questions can also be posed.

The general question – whether impacts from HEP are wholly site-specific or are generalisable into a set of principles – can be posed. In the literature, Trussart *et al.* (2002) indicate that each scheme is considered to be unique in its impacts.

3.9 Hydrology and hydraulics

In the fields of hydrology, the study of water on the earth's surface, and hydraulics, the science relating to the flow of fluids (Chambers 1974), concepts and techniques have been developed to explain and predict the behaviour of river flows in erosion and deposition, i.e. where the river's energy is directed as work. The subjects of hydrology and hydraulics are both extensive and highly specialised, and only certain aspects can be touched upon here.

The scope of this study does not extend to a comprehensive review of these fields, but uses some concepts and techniques from them. In particular, Manning's (1889) equations governing flow prediction in open and closed channels, and Bagnold's *stream power* (1966) concepts are relevant here. Stream power is one of the more important concepts, which has been widely used as a basis for river flow effects quantification and prediction.

3.9.1 Concept of stream power

A suitable already developed concept has been used in hydraulics to describe the power of streams, namely 'stream power'. Stream power expresses the power available to the river in watts, as a consequence of the specific force of gravity on water, expressed as a flow rate in m s^{-1}, and the slope or gradient expressed as a fraction. Stream power can be used in a variety of ways; per length of river reach termed 'total stream power', per metre of river termed 'cross-sectional stream power' or per defined unit of bed area. This concept, originated by Bagnold (1956, 1966), 'describes the force exerted by a mass of water moving over and across a single cross section per unit time', and is used by hydrologists (Fitzgerald and Bowden 2006). It is used to study sediment transport, bed-load movement, and to determine whether a stream is aggrading (accumulating sediment) or degrading (losing sediment or incising).

Equation: $\Omega = \rho g Q S$

(Bagnold 1966)

where:
Ω = stream power per unit measure of bed (W m^{-1})
ρ = density of water (kg m^{-3})
g = acceleration due to gravity (9.81 m^{-2})
Q = discharge (m^3 s^{-1})
S = slope (dimensionless)

The term 'stream power' has also been defined as the 'rate of energy supply at the channel bed which is available for overcoming friction and transporting sediments' (McEwen 1994: 359; cited in Barker 2004). Flowing water performs work as potential energy is converted into kinetic energy, and can be said to have the properties of mechanical power. When this occurs within a channel, most of the mechanical energy is dissipated as friction with the channel boundary. However, a portion is left over for eroding and transporting sediment, i.e. geomorphic work. The concept has been used widely in the literature to calculate sediment transport, to explain channel incision and channel pattern and bank-side habitat development.

General physics equations traditionally describe force per unit area when calculating mechanical power (for example, watts and horsepower), as has been done here in formulating this hypothesis, however the stream power concept uses the slope and gradient profile to calculate theoretical power. The apparent suitability of this concept for measuring and testing the hypothesis on HEP leads to the following question: could the use of the concept of stream power provide a means of testing whether the environmental impacts of HEP developments are proportional to power flux density?

Bibliography

Alexander, G. (1996) 'Overview', in Boyle, G., ed., *Renewable Energy*, Oxford University Press and the Open University, Oxford and Milton Keynes.

Bagnold, R.A. (1956) 'The Flow of Cohesional Grains in Fluids', *Royal Society London, Proceedings, Series A*, Vol. 249: 235–97.

Bagnold, R.A. (1966) 'An Approach to the Sediment Transport Problem from General Physics', USGS Professional Paper.

Barker, D. (2004) 'Modelling Downstream Change in River Flood Power: A Novel Approach Based on the UK Flood Estimation Handbook', University of Birmingham online resources: www.iem.bham.ac.uk/water/barker.htm, accessed 22 July 2009.

Bartle, A. (2002) 'Hydropower Potential and Development Activities', *Energy Policy* Vol. 30: 1231–9.

Betz, A. (1920) 'Das Maximum der theoretisch möglichen Ausnutzung des Windes durch Windmotoren', *Zeitschrift für das gesamte Turbinewesen*, Vol. 26: 307–9.

Betz, A. (1966) *Introduction to the Theory of Flow Machines*, trans. Randall, D.G., Pergamon Press, Oxford.

Boyle, G., ed. (1996) *Renewable Energy Power for a Sustainable Future*, Oxford University Press in conjunction with the Open University, Milton Keynes.

Brandon, M. and Smith, S. (2003) 'Water', in Morris D., Freeland J.R., Hinchcliffe S.J. and Smith S., eds, *Changing Environments*, John Wiley and the Open University, Chichester.

Catlow, J. and Thirlwall, C. (1976) *Environmental Impact Analysis*, Department of the Environment Research Report II, HMSO, London.

Chambers (1974) 'Dictionary of Science and Technology', W & R Chambers, Edinburgh.

Charlier, R.H. (2003) 'Sustainable Co-Generation from the Tides: A Review', *Renewable and Sustainable Energy Reviews*, Vol. 7: 187–213.

Clarke, A. (1993) 'Comparing the Impacts of Renewables: A Preliminary Analysis', TPG Occasional Paper Technology Policy Group 23, Open University, Milton Keynes.

Clausius, R. (1856) *The Mechanical Theory of Heat with Its Applications to the Steam Engine and to Physical Properties of Bodies*, John van Voorst, London.

ETSU (1994) *An Assessment of Renewable Energy for the UK*, HMSO, London.

Fitzgerald, E. and Bowden, B. (2006) 'Quantifying Increases in Streampower and Energy, Using Flow Duration Curves to Depict Stream Flow Values', *Stormwater*, Vol. 7 No. 2: 88–94.

Fraenkel, P. (2007) 'Marine Current Turbines: An Update', MCT Ltd online resources: www.all-energy.co.uk/UserFiles/File/2007PeterFraenkel.pdf, accessed 18 November 2008.

Freer, R. (2001) 'The Three Gorges Project on the Yangtze River in China', *Proceedings of ICE, Civil Engineering*, Vol. 144: 20–8.

Gagnon, L., Belanger, C. and Uchiyama, Y. (2002) 'Life Cycle Assessment of Electricity Generation Options: The Status of Research in Year 2001', *Energy Policy*, Vol. 30: 1267–78.

IEA (2014) 'Key World Energy Statistics', International Energy Agency, online resources: www.iea.org/publications/freepublications/publication/KeyWorld2014.pdf, accessed 20 May 2015.

Key, M.H., Hutcheon, R.J., Preston, D.A. and Donaldson, T.P. (1972) 'Ultraviolet and X-ray Spectroscopy of Astrophysical and Laboratory Plasmas', *Space and Science Review*, Vol. 13: 584.

MacKay, D. (2008a) 'Technical Chapters to Sustainable Energy', online resources, http:// www. withoutthehotair.com/cft.pdf, accessed 13 May 2016.

MacKay, D. (2008b) *Sustainable Energy Without the Hot Air*, Cambridge UIT, online resources: www.inference.phy.cam.ac.uk/sustainable/book/tex/cft.pdf, accessed 20 July 2009.

Manning, R. (1889) 'On the Flow of Open Water in Channels and Pipes'. *Transactions of the Institution of Civil Engineers of Ireland*, Vol. 20: 161–6.

McEwen, L. (1994) 'Channel Planform Adjustment and Stream Power Variations on the Middle River Coe, Western Grampian Highlands, Scotland'. *CATENA*, Vol. 21 No. 4: 357–74.

Ong, S., *et al.* (2013) 'Land Use Requirements for Solar Plants in the United States', Technical Report NREL/TP-6A20–56290 June 2013, National Renewable Energy Laboratory, online resources: www.nrel.gov/docs/fy13osti/56290.pdf, accessed 19 September 2015.

Petts, G.E. (1984) *Ecology of Impounded Rivers: Perspectives for Ecological Management*, John Wiley & Sons, Chichester.

Ramage, J. (1996) 'Hydroelectricity', in Boyle, G., ed., *Renewable Energy Power for a Sustainable Future*, Oxford University Press in conjunction with the Open University, Milton Keynes.

Ramage, J. and Scurlock, J. (1996) 'Biomass', in Boyle G., ed., *Renewable Energy Power for a Sustainable Future*, Oxford University Press in conjunction with the Open University, Milton Keynes.

Trussart, S., Messier, D., Roquest, V. and Akil, S. (2002) 'Hydropower Projects: A Review of Most Effective Mitigation Measures', *Energy Policy*, Vol. 30: 1251–9.

Twidell, J. and Weir, A. (1986) *Renewable Energy Resources*, E&FN Spon, London.

WEC (2004) *Comparison of Energy Systems Using Life Cycle Assessment*, World Energy Council, London.

4 Questions raised by the hypothesis

4.1 Introduction

The central question for the research described in this book is: *Is the environmental impact of renewable energy sources related to the power flux density?* The hypothesis is that a relationship exists, but that it may not be a simple one in all respects.

The conclusions to the last chapter outlined the proposed parameters of the hypothesis. These parameters raise some questions.

Relating directly to power flux density:

- Does higher power flux density of a natural energy flow result in a higher rate of work of functions performed?
- Does higher power flux density result in lower land use per unit of energy?

Relating to the proportion of energy abstracted from a renewable energy flow:

- Do impacts result from the reduced energy available for natural functions?
- Is the degree of impact related to the proportion of energy extracted from a natural energy flow?

Relating to efficiency of conversion by the converter device:

- Does the efficiency of energy conversion affect the proportion that can be abstracted?
- Does the efficiency of energy conversion affect the production of unwanted by-products?

Relating to the number of energy conversions:

- Are the impacts proportional to the number of conversions to the form of the energy flow?

These questions can be further broken down into the different forms of impact, the different renewable energy sources, and the type of relationship with power flux density.

4.2 Central question formulation applied to HEP

The central question of this book – is the environmental impact from renewable energy sources related to the power flux density? – could be applied to the test case of HEP in a number of ways.

As stated earlier, the first and most apparent way to show this could be through the land use or area required by different renewable energy sources, or more particularly HEP. Land use can then be equated with environmental impact; the greater the use of land, the greater the impact. For example, Egre and Milewski (2002) cite the area inundated as a rule of thumb measure of impact for HEP schemes.

This should be relatively straightforward to show, albeit with some complications such as the land area actually occupied as opposed to land affected but not directly occupied.

A question arising from HEP as a test case would be whether impact increases in proportion to the size of development or not. That is, does impact scale linearly or non-linearly as the power flux density increases? This could be considered for HEP, where one large scheme with high head could be compared with a series of smaller schemes with lower heads.

4.2.1 Is there a relationship between the power flux density and functions in the natural environment?

The hypothesis proposes that the power flux density of any renewable energy source may characterise the environmental impacts, due in part to the reduced energy available for natural functions after extraction and in part due to inefficiencies of conversion and changes in the work performed per unit area. This central question is based on the idea that natural energy sources power natural processes in the environment. Another way of expressing this is that a natural energy flow performs a number of *functions* in the environment. These functions could be important to the maintenance of any environment. For example, river systems are responsible for erosion, transport and deposition of sediment (Petts 1984). Such processes can maintain bank shapes and deltas, which would otherwise be eroded by the action of waves and sea currents. Further general functions of natural energy flows were given in Table 1.1.

A number of basic questions can be asked. Following on from the theory in Chapter 3, but in more detail:

- Is it true that the greater the power flux density, the greater the number and intensity of the functions performed by the flow?

- Does a lower power flux density result in fewer functions being performed by the flow?
- If there is a relationship, is it linear or geometric, exponential or some other?
- What limits might apply to this relationship?

4.2.2 Discussion on how impacts are to be measured

For the test of the hypothesis, the question of how the impacts are to be measured in the case of HEP can be posed. Impacts could be measured quantitatively from the data of individual impact categories, where available, and qualitatively through literature and possibly case studies.

Specifically for the test of the hypothesis, the primary questions posed are:

- What is the power flux density?
- Is there any link between this power flux density and the impacts, in particular sediment transport and land use?
- What proportion of the flow is intercepted? (i.e. the interception ratio).
- What proportion of the energy of the flow is abstracted?

Applying the parameter of proportion of energy flow abstracted to rivers and HEP schemes poses certain problems. For example, the straightforward abstraction of energy from a flow can be considered, and hence a reduction in flow energy downstream would occur. Alternatively, the flow might be substantially changed and modified in the process of abstracting a proportion of the energy, in such a manner that the whole of the flow had been affected or perturbed although only a proportion of the energy had been extracted.

In the case of HEP schemes, dams can either intercept the whole of the river bed and thus the whole of the flow, or alternatively, in the case of smaller schemes, smaller dams or weirs might be used to allow a leat (an almost-level channel) to be run along the valley side, taking a proportion of the river flow, to achieve sufficient head height further along as the valley drops away.

In this latter case the abstraction of the water flow away from the main river channel is manifest, and the energy abstracted could be calculated. However, most large HEP schemes have dams which intercept the entire valley, and thus the entire river flow, although not all of the river's energy that has been concentrated at the dam will be converted to electricity and abstracted; some of it may still be available to the river, albeit in an altered form.

HEP schemes pass some of the flow over the spillways or around the dam, and only a proportion of the flow is routed via the sluices and penstock to the turbines to generate electricity. This may, in the case of some large schemes, be a high proportion, or in others a lower proportion.

4.3 Flood erosion: alluvial channels

Baker *et al.* (1988) state that

> Erosional effects in alluvial channels do not correlate directly with hydraulics, for example, mean velocity or discharge is a consequence of regime behaviour. Adjustments in sediment concentration and bed roughness are the variables that are most difficult to evaluate in predicting alluvial channel erosion. The only effective procedure derives from the tendency of alluvial rivers (a) to conserve adjustments that lead to equilibrium and (b) to dissipate adjustments that do not.
>
> (Maddock 1976; cited in Baker *et al.* 1988)

This observation on regime expressions suggests that in terms of functions, the different forms that the power flux density can be applied to under different river bed conditions (e.g. shallow, broad beds, as opposed to deep, narrow beds), merely alter the type of function (e.g. erosion, deposition or increased flow velocity), not the overall magnitude of the generalised effect. In other words it may be of no consequence (to the hypothesis) that a river is eroding, depositing or increasing/decreasing sediment transport, as all of these qualify as functions. These actions are components of a larger process, i.e. the tendency of rivers to shape landscapes, smoothening and flattening, but in a continuum through the balanced equilibrium of the three processes: erosion, transport and deposition. The regime equations by Baker *et al.* (1988) describe some of the dynamics of equilibrium states. Each equilibrium state then has a threshold value at which it will change to a new equilibrium value.

4.4 Which impacts are unrelated to energy?

Since the purpose here is to relate impacts to the power flux density and energy flows, it is worth identifying some of the impacts which are not related directly to energy. For example, it could be argued that the extent of population displacement by a hydro electric scheme depends on the population density of the area, rather than an energy-related factor. This is partly true, as the example of the Three Gorges HEP scheme in China demonstrates (Ziyun 1994). The high population density of the valleys affected by the reservoir has indeed resulted in high displacement. However, another factor contributing to this is the length of the reservoir, at 600 km (Greeman 1996), which in itself is due to the low gradient of the river and the head height. So while a non-energy-related factor is the main reason for the large population displacement, energy-related factors also play a role.

Rare or valued wild plant or animal populations in the area endangered by inundation by the reservoir might also constitute impacts that are unrelated to energy. There have been examples of endangered species being threatened by HEP reservoirs. However, once again the extent of land used per unit of energy harnessed, which could relate to power flux density, could be a significant factor. A point worth noting here is that river valleys can be rich in biodiversity and some, such as the Three Gorges, have provided habitats for rare species (Ziyun 1994), partly due to their micro-climates or unique conditions. These may be partly energy related. To some extent, then, some impacts could be described as 'accidents of geography', and be relatively unrelated to energy factors.

Another impact not directly related to energy is that of the barrier to migratory fish such as salmon and trout that are formed by dams and weirs for HEP schemes. Fish passes can be designed to by-pass such obstacles and the turbine in-take and outlet. Crucially, the efficacy of fish ladders depends on water flows being adequate to attract the fish rather than the flows at the weir or turbine duct (Environment Agency 2008). Minimum flows need to be established during the spawning period, even if this means less flow via turbines and hence reduced generation. This illustrates an example of the concept of proportion of flow, applied to maintaining conditions in a river. However, some rivers, such as the Columbia, have supported larger fish populations than others (Lang 2007).

Impacts from construction might be considered to be unrelated to energy. Apart from the impact of erecting large structures, where the bigger the structure the bigger the impact might be expected to be, there may well be ways of reducing the impact of construction which are unrelated to energy. One example here is the use of a coffer dam for a rock barrage when building the Rance tidal barrage in France, as opposed to modern techniques of floating in caissons, which can lower the impact significantly (Baker *et al.* 1988). Impacts from construction such as noise, fumes, dust or debris relate more to the techniques employed as opposed to any energy factor.

4.5 Wider relevance of the hypothesis to other renewable sources

The question then arises whether the hypothesis applies to other renewable energy sources apart from HEP. Do the parameters identified earlier, i.e. *power flux density*, *proportion of the flow intercepted and energy abstracted*, efficiency of conversion and *number of conversions* apply to environmental impact elsewhere, for example to tidal energy, or wave energy, or moving further from HEP, to wind, solar and biomass sources?

These sources differ in a variety of ways, but can be ranked in terms of power flux density. Is this parameter then related to impacts and in what manner?

Since these sources diverge increasingly in their principles of operation from the starting position of HEP, can their flows be assessed in a similar manner? If this is possible, are the proportion of the flows intercepted and the proportion of the energy abstracted significant in the impact caused? Whether or not this relationship can be demonstrated for all sources of renewable energy constitutes a further question.

Table 4.1 General levels of impact from diffuse or dense renewable flows on environmental categories

Environment category	*Ratio of energy abstracted*			
	Diffuse flow		*Dense flow*	
	Low	*High*	*Low*	*High*
Human	Medium	Big	Small	Small
Animate	Small	Big	Medium?	Big
Inanimate	Small	Medium	Medium	Big

Source: Clarke (1993).

The efficiency of energy conversion parameter of devices and processes can readily be applied to the different renewable energy sources; its links to impacts and whether this can be shown is a separate question.

Related to the efficiency of energy conversions is the number of conversions that occur in the converter device, which should be identifiable, through close description of the processes and principles involved. To what extent these energy conversions account for some of the environmental effects through, for example, changes or discontinuities in the functions in nature of those energy flows, is another question.

In Clarke (1993), the author suggested that the relevance of the hypothesis lay in part in the characterisation of impacts from renewables, such that impacts did not just increase proportionately to the amount of development, but depended on the power flux density of the source flow as well as the proportion of energy abstracted, and also on the category of the environment.

A summary table outlining the expected impact level from either diffuse or dense flows on different environmental categories is shown above in Table 4.1. Until tested this must remain hypothetical, though the pattern of impact placement is consistent with the reasoning provided above.

4.6 Conclusions

The central questions concern the power flux density:

- whether a higher power flux density results in a higher rate of functions performed;
- whether less land is used per unit of energy by higher power flux density sources

Next, questions concerning the proportion of the energy flow abstracted or energy extracted were posed:

- whether impacts result from reduced energy available for natural functions;
- whether the degree of impact relates to the proportion of energy extracted from a natural energy flow.

Following on, questions concerning the effects of efficiency of conversion were posed:

- whether the efficiency of energy conversion affects the proportion that can be extracted;
- whether the efficiency of energy conversion affects the production of unwanted by-products.

Finally, questions concerning the number of energy conversions of the flow were posed:

- whether the number of energy conversions is reflected in the impacts.

These primary questions raised a number of other issues concerning methods of measurement and how the relationships would be identified. The requirement for a single source to be chosen as a test case was discussed. HEP was chosen due to its maturity, long experience and widespread application, as well as availability of data.

The applicability of the parameters and questions posed to HEP were considered, for example in regard to power flux density and land use. The issue of how impacts from HEP increase with scale, whether it is linear or non-linear was raised and there was discussion of how the functions performed by natural energy flows were possibly related to power flux density. The manner in which natural energy flows maintain natural environments was illustrated using the example of river flows. Energy flows resulting in a balance of forces maintain the environment, in this case the river, in a state of dynamic equilibrium. Impacts can then be defined as changes that push a stable system into a new equilibrium state. Existing processes of slow stable change, which are interrupted, reversed or terminated, would then be considered as impacts.

Some considerations in applying the concept of proportion diverted to HEP developments were considered, for example the need to distinguish between diverting water, changing the nature of the flow, perturbing it and actually extracting energy from the flow. In most cases of large HEP development, the entire valley is dammed and the entire flow intercepted, with resulting changes. However, some schemes do not intercept the whole flow. The actual proportion of energy extracted is a related but distinct factor.

The relevance of analysis techniques used in the subject area of hydraulics was considered. Two techniques that have relevance here are Manning's equations for open channel flows and Bagnold's stream power concept. Some regime expressions for flood erosion in alluvial channels are outlined.

Those impacts not directly related to energy were discussed; for example, population displacement from HEP schemes, aesthetic reactions to wind energy and biodiversity issues for biomass energy sources. Accidents of geography may result in particular impacts, but others are unrelated, e.g. manufacturing impacts.

The question of the wider relevance of the hypothesis to other renewable energy sources, as well as HEP, might be considered, since other sources such as tidal, wave, solar, wind or biomass differ from HEP in such respects as power flux density and the nature of their flows. This is left to later chapters.

The suggestion that impacts arise not just from the size of developments, but also depending on the nature of the energy flow as well as the category of the environment, was considered. Impacts from diffuse or dense flows may be *experienced disproportionately by certain environmental categories.*

A number of further questions raised by the hypothesis that cannot all be answered in this study but could form part of further research are included in the conclusions *in Chapter 9.*

Bibliography

Bagnold, R. A. (1956), 'Flow of Cohesionless Grains in Fluids': Royal Society of London, Philosophical Transactions,Vol. 249: 239–57.

Bagnold, R.A. (1966) 'An Approach to the Sediment Transport Problem from General Physics', USGS, Professional Paper.

Baker, V., Kochel, R. and Patton, P. (1988) Flood Geomorphology John Wiley & So*ns, Chichester.*

Barker, D. (2004) 'Modelling Downstream Change in River Flood Power: A Novel Approach Based on the UK Flood Estimation Handbook', University of Birmingham online resources: www.iem.bham.ac.uk/water/barker.htm 2004-2007, accessed 22 July 2009.

Bartle, *A. (2002) 'Hy*dropower Potential and Development Activities', *Energy Policy, November.*

Ca*t*low, J. and Thirlwall, C. (1976) Environmental Impact Analysi*s, Department of Environment Researc*h, HMSO.

Chambers (1974) Dictionary of Science and Technology, W&R Chambers, Edinburgh.

Clarke, A. (1993) 'Comparing the Impacts of Renewables: A Preliminary Analysis', TPG Occasional Paper Technology Policy Group 23, Open University, Milton Keynes.

Egre, D. and M*ilewski, J.C.* (2002). 'The Diversity of Hydropower Projects'. Energy Policy,Vol. 30 No. 14: 1225–30.

Environment Agency (2008) 'History of the Dee Scheme', Environment Agency, online resources: www.environment-agency.gov.uk/homeandleisure/drought/38583.aspx, accessed 15 December 2008.

Fitzgerald, E. and Bowden, B. (2006) 'Quantifying Increases in Streampower and Energy, Using Flow Duration Curves to Depict Stream Flow Values', *Stormwater*, Vol. 7 No. 2: 88–94.

Freer, R. (2001) 'The Three Gorges Project on the Yangtze River in China', *Proceedings of ICE, Civil Engineering*, Vol. 144: 20–8.

Greeman, A. (1996) 'Yoking the Yangtse', *New Civil Engineer*, 14 November.

IEA (2006) 'Renewables in Global Energy Supply', IEA Fact Sheet, September, online resources: www.iea.org/textbase/papers/2006/renewable_factsheet.pdf, accessed 7 August 2009.

International Water Power and Dam Construction (1996) *Year Book*, Reed Business Publishing, London.

Lang, W. (2007) 'Columbia River History', Portland State University, online resources: www.ccrh.org/river/history.htm, accessed 18 October 2007.

McEwen, L. (1994) 'Channel Planform Adjustment and Stream Power Variations on the Middle River Coe, Western Grampian Highlands, Scotland.' *CATENA*, Vol. 21 No. 4: 357–74.

Manning, R. (1889) 'On the Flow of Open Water in Channels and Pipes', *Transactions of the Institution of Civil Engineers of Ireland*, Vol. 20: 161–6.

Petts, G.E. (1984) *Ecology of Impounded Rivers: Perspectives for Ecological Management*, John Wiley & Sons, Chichester.

Ramage, J. (1996) 'Hydroelectricity', in Boyle G., ed., *Renewable Energy Power for a Sustainable Future*, Oxford University Press in conjunction with the Open University, Milton Keynes.

Trussart, S., Messier, D., Roquest, V. and Akil, S. (2002) 'Hydropower Projects: A Review of Most Effective Mitigation Measures', *Energy Policy* November.

Water Power and Dam Construction (2007), 'Beyond Gorges in China'. Online resources: www.waterpowermagazine.com/news/newsbeyond-three-gorges-in-china, accessed 26 January 2016.

Ziyun, F. (1994) *Environmental Issues of the Three Gorges Project*, International Water Power & Dam Construction, London.

5 Constructing the test

The previous chapter has described the main questions raised by the theory, in particular as applied to hydro electric power (HEP). The central question of whether a higher power flux density results in a higher rate of work (i.e. water flow eroding and transporting sediment) in relation to functions performed by the natural energy flow was posed. Further questions raised were whether a higher power flux density source uses less land area per unit and whether reduced energy available for natural functions leads to impacts. Moreover, the questions of whether these impacts were in proportion to the energy extracted, whether the efficiency of conversion is a factor in impacts, and also whether this was influenced by the number of energy form conversions, were asked. This chapter describes the setting up of the test of the hypothesis by exploring these questions in relation to HEP.

5.1 Introduction

This chapter investigates the issues and feasibility of designing a test to investigate in broad terms the environmental impact associated with HEP. Some of the impacts of HEP schemes have been outlined and are briefly discussed below, as the author has reviewed these in Clarke (1995). In order to test the hypothesis, initially a set of hydro electric schemes which have become known for environmental effects were selected and their dam and reservoir parameters were researched, to examine whether these could be linked to impacts. Following this, the river courses below the hydro electric schemes in question were investigated in terms of the energy available, and gradient profiles were plotted for four of the rivers, where data were available. The concept of total stream power was applied to four of the rivers – the Danube, the Nile, the Columbia and the Colorado. Where data were available, the available energy in the river before and after the HEP scheme was modelled in coarse form. This loss of energy, and in particular the power flux density, could then be tested for links to impacts downstream of the scheme. The environmental effects associated with HEP have been generally described, as well as those linked to specific schemes, and there are descriptions of the four individual river systems which have been considered in more detail.

5.1.1 General impacts of hydro electricity

Hydro electric power has existed for over a century (Ramage 1996), and is the largest source of renewable electricity, supplying ~16.2 per cent (IEA 2014) of world electricity. The world's largest power stations are hydro electric, for example, the 22.5 GW Chinese Three Gorges scheme or the 12 GW Brazilian & Paraguayan Itaipu scheme (International Water Power & Dam Construction 1996), though HEP can be developed at almost any scale, from ~1 kW to 22 GW, provided the resource is available. It is considered to be a mature technology which is well known and understood. HEP schemes, especially large ones, are acknowledged to cause substantial change to rivers which result in a wide variety of environmental effects and impacts (Petts 1984); see Table 5.1.

There is a considerable body of literature on its environmental impacts. All HEP schemes will have some effects, since energy is abstracted from the

Table 5.1 Summary of general impacts from different types of hydro electric power scheme

High-head dam and lake	*Low head/run of river/weir*
Lacustrine environment	
Loss of river gradient	Smaller change to river gradient
Land Use	Small/no land use
Population displacement	Little/no population displacement
Incompatibilities	Few/no incompatibilities
Reduced water speed	Smaller speed reduction
Wildlife flora and fauna change	Smaller changes to flora and fauna
Silting	Smaller silting
Nutrient loss downstream	Little nutrient loss
Water quality reduction	Lower water quality reduction
Supersaturation possible	Supersaturation unlikely
Eutrophication	Eutrophication unlikely
Reduced species diversity possible	Species diversity reduction unlikely
Changed species	Species little changed
Water table effects	Little/no water table effect
Drainage effects	Fewer drainage effects
Flow changes downstream	Less flow change downstream
Increased erosion downstream	Less erosion downstream
Water temperature changes	Water temperature changes unlikely
Fish migration barrier	Reduced obstacle to fish migration
Fish turbine strikes	Reduced risk of fish turbine strikes
Fish damage from pressure changes	Pressure change fish damage less likely
Flood protection (+)	Less flood protection
Flow regulation	Less flow regulation
Often raised levels for navigation	Raised levels for navigation

Source: Clarke (1995)

flowing stream, and changes are made to the channel by dams, and in many cases water is diverted (Petts 1984). Flow changes downstream can result from reservoirs, which can affect the transport of sediment. Most commentators believe the impacts downstream from the dam to be more serious than those at the reservoir and dam reach (Petts 1984). 'The most serious problems resulting from the construction of hydropower facilities are the disruption of the longitudinal continuity of the rivers, and dramatic changes in the rivers' hydrological characteristics' (Joint Danube Survey 2005). The present author has written on this topic – see Table 5.1 of general impacts that HEP schemes can result in.

5.1.2 Selection of case sample

A selection of environmentally significant HEP plant was made, based on some of the best-known examples around the world. These are all large schemes, based on seven different major rivers, in five continents: the Gabcikovo and Iron Gates schemes on the Danube River in Central and Eastern Europe; the Columbia River schemes in the north-west USA; the Colorado River's Hoover and Glen Canyon schemes in the south-west USA; the Itaipu scheme on the River Parana; and in Africa, the Aswan High dam on the Nile, and the Kariba and Cabora Bassa schemes on the Zambezi; and finally the Three Gorges scheme on the Yangtze in China. A list of these schemes is shown below in Table 5.2.

Of these HEP schemes, it is probably fair to describe the Aswan High scheme as the best known for its environmental impacts downstream. The Nile is one of the best-studied rivers in the world, and together with the Danube there are good flow records going back to the nineteenth century.

5.1.3 Definition of environmentally significant dams

Environmental significance is defined here as due to the number and severity of environmental impacts as reflected in literature citations, such as the

Table 5.2 List of HEP schemes in this study

- Aswan High
- Kariba
- Cabora Bassa
- Itaipu
- Hoover Dam
- Glen Canyon
- Bonneville Dam
- Grand Coulee
- Gabcikovo
- Iron Gates I & II
- Three Gorges

number of scientific journal papers, or the number of Google citations. Controversy and political disputes are other signifiers of notoriety, as are campaigns by non-governmental organisations (NGOs). The impacts cited and their severity provide further indications. While this definition is not an objective one, it should, to some extent, reflect the sensitivity of perceived environmental impacts, socially and in terms of the physical environment. This sensitivity to impacts is likely to have had the effect of inducing greater scrutiny, resulting in more research being carried out and more publications.

5.2 The river systems characterised

In attempting to compare dams and HEP plants on different river systems, the individual and unique characteristics of each river system need to be taken into account. Each of the seven different rivers considered here flows in different latitudes, climates, terrains and geology. Each has its own flow characteristics, e.g. fairly regular seasonal flow variations, reflecting the climate and precipitation in its catchment area, as well as soil and geological rock types which influence run-off and flow. Each river system has its own individual gradient pattern, as well as catchment vegetation (that is, biomes), due to latitude, climate and altitude. For each river a unique human population settlement pattern will exist, as also will populations of flora and fauna. In short, each river system has a unique combination of factors making up its particular environment. This will result in particular individual sensitivities. Despite this apparent site specificity, it is proposed that there could be enough common factors to enable different rivers to be compared by using common parameters. However, the unique characteristics of each river system need to be identified and then the data set needs to be normalised. They are described in the following section for the four rivers whose gradient and energy profile has been plotted.

5.2.1 Nile

The Nile is (arguably), the world's longest river with a length of 5,611 km, not including the river Kagera section, the largest tributary of Lake Victoria. The University of New Hampshire/Global Run-off Data Centre UNH/GRDC flow station networks cite the total length of the main stem as 5,964 km, to basin outlet (UNH GRDC 2007). The Nile has been thoroughly surveyed and studied and records of its flows extend back to 3000 BC in some form (Sutcliffe and Parks 1999). Records have been kept at Aswan since 1869, and although not entirely continuous and consistent, are fairly complete compared to other major rivers. The river's flow direction is south to north (northwards) from a latitude of 3°S to 32°N, from south of the Equator to the Mediterranean, though its catchment traverses only 8° of longitude. Its theoretical drainage area is three million square kilometres, though of this about 44 per cent does not contribute any rainfall (Rzoska

1978). Of all the world's major rivers, the Nile has the lowest specific mean discharge per area of catchment, at 0.98 m^3 km^{-2} (Marndouh 1985). The Nile descends from a height of 1,134 m on the East African plateau, to 457 m on the Sudan plains, over 800 km, and then drops to 372 m by Khartoum, flowing on for 3,000 km through desert and a series of rocky sills known as cataracts, to reach the Mediterranean at a great delta. Below Cairo, the river splits into two main branches, the Damietta and Rosetta. The Nile's unusual catchment shape is thought to be the result of a merging of several separate river systems in the final phase of the Pleistocene. This was caused by the tectonic tilting of the East African plateau, which contributed to the formation of Lake Victoria. Large-scale faulting has formed the Rift Valley and affected the direction of the Nile.

The Blue Nile, the main tributary and its other two main tributaries, the Sobat and the Atbara, flow down from the volcanic high Plateau of Ethiopia, probably formed in the Oligocene, about 20 million years ago. Evidence from sediments indicates that these rivers flowed north independently from upper parts of the river and only joined up relatively recently (Rzoska 1978). The Blue Nile produces much of the flow and most of the flood flow, as its catchment area has one wet season. Being a steep river with its flow unattenuated by the lakes which occur on the White Nile, it also contributes the majority of the crucial sediment load, estimated as 140 m tons per year (Sutcliffe and Parks 1999).

The shape of the Nile basin has been caused largely by the slopes of the rivers, with layers of alluvial soils spread across central Sudan, resulting from Nile floods. The Nile has been heavily influenced by climatic changes over the last 30,000 years, with considerable shifts of rainfall, with a series of wet and dry phases, which have affected North Africa and Egypt and the present Sahara region. Especially over the last 20,000 years, changes of climate and vegetation have occurred.

There are a great variety of biomes, i.e. of rainfall and temperature and the resulting vegetation, over the great latitudinal extent of the river. Rainfall varies from 1500 mm per year at Lake Victoria, with two annual peaks, to 25 mm and then nil over the last 1,500 km through Sudan and Egypt. The East African lake plateau has a temperature of ~25 °C throughout the year and vegetation is mixed savannah and woodland. Ethiopia was originally covered in mountain vegetation, now very altered by man. The Blue Nile gorge has a unique habitat with specific climatic zones, influenced by successive flooding and aridity, with fringe forest and scrub. In Sudan the river environs have belts of lowland forest, with small areas of montane vegetation, followed by savannah woodlands, swamp, and wetland savannah, then thorn savannah and finally semi desert and desert, as a result of the diminishing rainfall downstream (Sutcliffe and Parks 1999). The alluvial cultivated valley of the Nile in Egypt and that formerly in Nubia, are surrounded by desert. The narrow river valley widens out into the 'great riverain oasis of Faiyum' and then the delta (Sutcliffe and Parks 1999).

Dams on the Nile

Since ancient times, barrages have been used to intercept the Nile's waters for irrigation purposes. Today there are six major dams on the Nile, with only the Aswan High dam being of the impoundment type. Excluding the barrages for irrigation, and the first Aswan dam of 1904, the dams on the Nile are the Owen Falls dam in Uganda, at the exit to Lake Victoria, used mainly for electricity generation, which raises the height of Lake Victoria slightly, by 3 m (Marndouh 1985); the Roseires dam of 1966 in Sudan, where the Blue Nile enters the White Nile; the Gebel Aulia dam of 1936, which is on the White Nile; the Senner dam of 1925 on the Blue Nile in Sudan; and the most recent, the Merowe dam of 2009 in Sudan. The Aswan High dam, of 1964, is used for electricity generation, irrigation and water supply, and the reservoir hosts a fishery. A former dam still exists at Aswan, built by the British government in 1904 (Rzoska 1978).

Below this on the Nile are a number of weirs and barrages associated both with navigation with locks and water supply for irrigation and for domestic and industrial consumption. These are the Esna barrage, the Nag Hammadi barrage and the Assuit barrage, and finally the Delta barrage and then the Edfina barrage (Marndouh 1985; Dickinson and Wedgwood 1982). Between the Aswan dam and Cairo the river can thus be divided into four reaches.

Sensitivity of the Nile's environment

Two countries, Sudan and Egypt, both of which have desert climates, rely almost completely on the waters of the Nile, since rainfall is either very low, in North Sudan, or almost completely absent in large parts of that country and in Egypt. Irrigation has been practised for thousands of years and, over time, river flows have been reduced by water abstraction for irrigation and water supply, as well as increased evaporation losses from wider channels and reservoir surfaces (Sutcliffe and Parks 1999). These losses are estimated to have increased from 2.2×10^9 m^3 to 3.2×10^9 m^3 in the period of records 1911–95 for the Atbara and main Nile between Khartoum and Wadi Halfa (Sutcliffe and Parks 1999).

It is also the silt and sediment washed down by the Nile that forms all of the useable soil in Egypt, and most in Sudan. In Egypt a narrow corridor of fertile land on each bank of the river, much of which is only 16–32 km wide, comprises the only useful agricultural land, until it widens out into the delta, below Cairo. The annual flooding of the Nile in Egypt was used in ancient times to determine the tax levied, as the amount of silt, and degree of soaking of the fields on either side of the river, determined the size of the harvest (Marndouh 1985). The delta itself was formed by the deposition of silt and sediments brought by the river.

The particular sensitivity of the environment concerns, first, the water supply, which is a vital resource for the millions of Egypt's population, and

Table 5.3 Impacts linked to the Aswan High Dam

- Sedimentation in the reservoir.
- Loss of sediment below the dam.
- Evening out of flow downstream of the dam, losing the flood discharge and sediment transport.
- Erosion of the delta.
- Loss of water by evaporation from Lake Nasser.
- Land area lost to the reservoir.
- Population displacement.
- Sedimentation and formation of a delta at the mouth of Lake Nasser.
- Fishery impact: loss of sardine fishery off the delta coast.

second, the silt and sediments which brought nutrient-rich replenishment to the country's soils. The delta is dependent on silt replenishment to replace coastal erosion by the sea. Since 1964, some 98 per cent of the 125 m tons (El-Moattassem 1994) of sediment transported by the river has been trapped in Lake Nasser, which has led to the near cessation of the replenishment of the delta. Impacts linked to the Aswan High dam are shown in Table 5.3.

However, since the Aswan High dam was built in 1964 and Lake Nasser formed, a constant all-year-round flow has been maintained on the river to supply water for domestic, industrial and particularly irrigation use, as compared with the seasonal drought and floods in September. In addition, an annual average of 8 TWh (Dubowski 1997) of electricity has been supplied, enabling much development in Egypt.

The normal pool level of the reservoir is 175 m, and the maximum is 183 m (El-Moattassem 1994). At 175 m the reservoir volume is 120×10^9 m^3, while at 183 m it is 164×10^9 m^3. Reservoir capacity comprises three parts, live storage of 90×10^9 m^3 between elevation 147 and 175 m, flood control capacity of 47×10^9 m^3 between 175 m and 183 m and dead storage capacity of 31.6×10^9 m^3 for use as a silt trap for 500 years (Amin 1994).

5.2.2 Danube

The Danube is Europe's second longest river, at 2,857 km, with a catchment area of 817,000 km^2 (Joint Danube Survey 2005: ch. 3). It rises at an altitude of ~690 m in the Black Forest in Germany, latitude ~48°N, and longitude ~8°E and flows generally eastward or east south-east, through Germany, Austria, Slovakia, Hungary, Serbia, Romania and Bulgaria, to reach the Black Sea at a delta in Romania, of which part borders the Ukraine, at latitude 45°N and longitude 27.7°E. Its major tributaries drain the North Eastern and Eastern Alps, e.g. the Lech, Isar, the Inn, the Drave, then south-west Czech Republic, i.e. the Morava, and much of Slovakia, the Vah, the Nitra and Ipel Ipoly. The Drave and the Save rivers drain, respectively, the Hungarian plain and Croatia and Serbia, while the Tisa flows south from the Carpathians through Hungary. The Morava River flows north, draining south-east Serbia and the border of Macedonia. A number of rivers flow south into the Danube from Transylvania in Romania, the Olt being one of the largest, along with the Jiul, the Vedea, the Arges and the Dambovita. The major tributaries, the Siret and the Prut, join within 250 km of the mouth, flowing from the north; the Prut forms the border between Romania and the Ukraine. As the river approaches the low reaches towards the delta it splits into several channels, with three main ones in the delta area.

The river was classified into three main sections by Laszloffy (1969; cited in Joint Danube Survey 2005); the upper is about 900 km long from source to the 'Hungarian Gate', where it breaks through the end of the Little Carpathians and drops down to the Hungarian Plain; then is the Carpathian Basin section of about 925 km, as far as the 'Iron Gate', where it breaks through the Southern Carpathians; the lower section follows at about 885 km, through Romania and Bulgaria to the delta. These three sections are characterised by their gradients from 1.1 per cent to 0.43 per cent in the upper section, to 0.17 per cent to 0.04 per cent in the middle, and 0.04 per cent to 0.01 per cent and 0.004 per cent near the mouth (Laszloffy 1967; cited in Joint Danube Survey 2005). The gradients are steeper at the Hungarian Gateway and the Iron Gates section, both having 'cataracts', which were pronounced at the Iron Gates. The Hungarian Gateway section is characterised by a braided stream, a steeper gradient and finally an inland gravel 'delta'. The gravel beds are very deep at this location. The Iron Gates consists of two sections of river gorge of steeper gradient for ~150 km, with great water depth; in the past there were submerged rocks and fast-flowing stream, now inundated by the two HEP scheme reservoirs.

The mean flow rate varies from 359 $m^3 s^{-1}$ in Germany at Oberndorf (GRDC 2008) to 1,463 $m^3 s^{-1}$ at Linz in Austria, and 2,047 $m^3 s^{-1}$ by Bratislava, to 2,354 $m^3 s^{-1}$ by Mohacs south of Budapest, and 5,500 $m^3 s^{-1}$ by the Iron Gates, and 6,415 $m^3 s^{-1}$ by Ceatal Izmail along one of the streams near the mouth. The hydrograph at Bratislava is characterised by a single peak in early summer in June of ~2,800 $m^3 s^{-1}$ from melted snow in the mountains, and a minimum of ~1,500 $m^3 s^{-1}$ in November. This seasonal variation is less pronounced downstream by the Nagymaros flow station. The river has been an important artery for Central and Eastern Europe for thousands of years, and is navigable for 2,600 km from Ulm (Joint Danube Survey 2005).

Dams on the Danube

There are 59 dams for hydro electricity along the upper Danube (Joint Danube Survey 2005) upstream of the Gabcikovo dam, as the steep gradient makes this suitable. There is a dam on average every 16 km on the upper section, and few reaches are free flowing. The largest HEP scheme is the Iron Gates I (1970) and II (1984) in Romania/Serbia; the second largest is the Gabcikovo in Slovakia. The Iron Gates I scheme has a head height of

26 m and Iron Gates II a head height of 7 m (Stankovic 1960), with the two reservoirs cited as extending some 270 km upstream (Joint Danube Survey 2005). Despite the relatively modest head height by comparison with the world's largest, the Iron Gates I is Europe's largest hydro electricity scheme, with a rated capacity of 2,136 MW due to the considerable flow rate.

Gabcikovo scheme

The Slovakian Gabcikovo HEP scheme, completed in 1992, was originally intended to be the first part of a dual-dam scheme, together with the Nagymaros dam in Hungary. The Nagymaros second dam was never built as the Hungarian government pulled out in 1989 on the basis of environmental objections at the time of democratisation (Borsos 1991). The Gabcikovo–Nagymaros scheme was intended as a 'peaking' power plant, and the second lower dam was intended to absorb the resulting wave from peak generation. Improvements to navigation were another prime objective, since the old course of the river was subject to low water in drought periods, limiting the draught of shipping over this reach.

The Gabcikovo dam in Slovakia is relatively unusual in that a huge 'leat' or artificially raised canal causeway takes water off 20 km from the main reservoir to provide the required head height, as the land and former main river stream fall away. Its head height is 21.5 m; with a mean flow of 2,047m^3 s^{-1} this provides for a rating of 720 MW. The Gabcikovo scheme diverts 80 per cent of the water flow into the 20 km leat (Zinke 2004), leaving the old main channel and the multiple braided channels either drought-prone or with sluggish water flows after weirs were installed, as opposed to the former dynamic flood flows. This area forms an inland 'delta' of deep gravel beds, where the steep gradient ceases at the entrance to the Hungarian Plains and contains protected wetlands (Zinke 2004). Another point of contention is that the old course of the river is also the international border between Slovakia and Hungary, and most of the water has been diverted away from the Hungarian bank. Slovakia has defended its continuation of the project alone, with some modifications, and has claimed that without the Nagymaros dam, about 2,000 GWh per year of electricity has been lost, worth six billion crowns, or about US$200 m (Liska 1993: 35). The unilateral decision to pull out of the scheme by Hungary provoked an international crisis which required third-party mediation. A summary of impacts for the Gabcikovo and Iron Gates I and II schemes is shown in Tables 5.4 and 5.5.

Neither of these schemes are impoundment dams and both have to be operated on a 'run of the river' basis; that is, they cannot store more than a small fraction of the river flow.

The remainder of the dams upstream have relatively low heads, and are run-of-the-river dams for hydro electricity and navigation purposes. According to the Joint Danube Survey (2005), about 30 per cent of the Danube's length is impounded for HEP schemes. Some 80 per cent of the river is regulated by

Table 5.4 Impacts linked to the Gabcikovo HEP scheme

- Loss of water flow in the old main channel.
- Loss of water supply to wetlands around the old main channel.
- Degradation of side channel flushing.
- Side channel siltation.
- Disruption of sediment transport.
- Ecosystem changes and degradation in the wetlands area due to drying out.
- Lowering of groundwater table in surrounding area due to diversion of main water flow (Zinke 2004).

Table 5.5 Impacts linked to the Iron Gates I and II HEP scheme

- Reduction of sediment transport.
- Bio-silicate deficiencies at the delta area.
- Nutrient changes at the Black Sea (Humborg 1997).
- Inundation of river banks.
- Settlement and population displacement.
- Reduction of sturgeon fisheries.
- Irrigation.

dykes or flood protection works, which have been built since the sixteenth century.

These include straightened sections, canals cut in the delta area and embankments cutting off the flood plains from the river. Only one-fifth of the former flood plain exists now compared to the nineteenth century.

The Danube delta

The Danube delta occupies an area of variously 3,446 km^2, 6,264 km^2 or 7,990 km^2 (WWF 2008) and is protected under the Biosphere UNESCO programme, the Ramsar Convention and also as a World Heritage Site, with 650 km^2 in Romania strictly protected (WWF 2008). The delta straddles the border between the Ukraine (20 per cent area) and Romania (80 per cent area). The Romanian protection dates from 1991; the Biosphere Reserve designation, shared between the two countries, dates from 1998. The Ukraine established the first protection in 1973 as part of the Black Sea State Reserve, then in 1981 as a Natural Reserve of the Danube stream.

The silt carried by the river resulted in an expansion each year of the delta, making it a dynamic region (Schwarz 2008). Recent commentators consider the delta overall to be declining in area. Schwarz states that only 34 per cent or 18 million tons p.a. of sediment reaches the delta, compared to 53 million tons p.a. before the construction of the Iron Gates dam (Schwarz 2008). Three main channels carry the waters to the sea and a multitude of small channels, though historically seven main channels existed. There are also two canals which cut across the delta, with a third one started by the Ukraine in 2004. This latter canal has caused controversy as it increases flows, resulting in less silt accumulation, and may have deleterious effects on the delicate ecosystem, which depends on water levels being maintained. The vegetation of the delta area consists of areas of forested islands that are regularly flooded, marsh areas and reed areas, with lagoons and lakes. It is an important bird roosting site, on migration routes for millions of birds, as well as a nesting and breeding site. There are 45 species of freshwater fish, 300 species of birds, and 1,200 species of plants (WWF 2008). Also, 15,000 inhabitants live in the delta area, many of whom live off fishing.

5.2.3 Colorado River

The Colorado rises at an altitude of 2,746 m in the Rocky Mountain National Park, and flows for 2,365.6 km through six different states, to reach the Gulf of California and the Pacific (Kammerer 1990). Its catchment area is approximately 626,780 km^2, and includes areas of Colorado, Utah, some of New Mexico, Arizona, Nevada and California, as well as Mexico. The river flows largely through desert areas, and its main flow is derived from melting snow in the Rocky Mountains, and some summer thunderstorms. For a large proportion of its length, it flows through canyons, carved from the arid rocks, the largest being the Grand Canyon at 446 km in length. Its average gradient over its whole length is one in 861, or 0.11608 per cent, though in its upper reaches it has steeper gradients (*Encyclopaedia Britannica* 2009).

The flow direction of the river is north-east–south-west and then north–south, from latitude 40°–32°N. The natural flow of the river has a very marked seasonal peak, and is quite variable, with a relatively low flow. Since the building of a number of large dam structures in the upper, middle and lower sections, the natural variability of the flow has been evened out by controlled discharges, with long water-storage retention periods. Lake Powell, the reservoir for the Glen Canyon dam, stores water for almost a year, and Lake Mead, the reservoir for the Hoover dam, for 2.8 years. Impounded volume of the reservoirs is about four times that of the river's average flow per annum (IBWC 2001: 33).

The International Boundary and Water Commission (IBWC) says that

> The operating criteria include three modes that govern releases from the dam. There is a minimum release of 8.23 million acre-feet (10.152×10^9 m^3) (per annum) to meet downstream demands. There are equalization releases to balance the amounts of water in Lake Powell and Lake Mead, though under certain conditions, equalization releases are not made. Additionally, spill avoidance is practiced whereby if high inflows are expected, water is released prematurely in order to create storage space.
>
> (IBWC 2001: 32)

The dams have multiple purposes, the primary ones being water supply for irrigation, flood control and hydro electricity supply. So much water is abstracted that the Colorado loses most of its flow by the time it reaches the Gulf of California, with just 23 $m^3 s^{-1}$ mean flow by 1975, at Yuma near the USA–Mexico border down from 875 $m^3 s^{-1}$ at Glen Canyon (USGS 2004). The major water supply is to the Imperial Valley for irrigation of citrus fruit groves, but water is also abstracted to supply the Las Vegas area and other regions. The irrigation canal, the 'All American Canal', takes off 450–850 $m^3 s^{-1}$, much of the river's flow in a normal rainfall year, to irrigate the Imperial Valley. Several major cities in the west, including Las Vegas, Los Angeles, San Diego, Tucson and Phoenix, rely either wholly or partly on the river for water supply, with aqueducts taking water off. About 90 per cent of the water of the river is used for irrigation and water supply.

Under the 1944 Water Treaty, the USA has an obligation to deliver 1.5 million acre-feet (or 1.8502×10^9 m^3) of water, ~10 per cent of river flow, to Mexico (IBWC 2001; Flessa *et al.* 2001). By the mid to late 1980s an average flow of 18 $m^3 s^{-1}$, and only 567.65×10^6 m^3 p.a., or ~30 per cent of this was delivered. Post-1991 to 2001, more water was supplied from the USA to Mexico, after recognition of issues surrounding water supply (Colorado River Delta Bi-National Symposium Proceedings 2001).

It has been noted that 'Most of the years, the river below the (international) border is dry' (Colorado River Delta Bi-National Symposium Proceedings 2001: 39)'About 20 percent of the total river flow in the past 20 years (since Lake Powell filled) have been flood flows and these typically come with El Niño events' (Colorado River Delta Bi-National Symposium Proceedings 2001: 46). 'In January 1997, there was a release of approximately 250,000 acre-feet of water over a three month period' (Colorado River Delta Bi-National Symposium Proceedings 2001: 46).

A population reduction of bi-valve mollusk from 34 to 95 per cent has taken place since diversion of the Colorado River water (Colorado River Delta Bi-National Symposium Proceedings 2001: 49).

> In 2001, about ten million acre-feet of water will be released from Hoover dam to meet the needs of downstream users. None of that water is released for ecological purposes and instead is used to meet contracts, agreements and the 1944 Water Treaty. Ten million acre-feet is one million acre-feet more than was needed to be released five years ago. Uses in the U.S. continue to grow and the states continue to use more of their entitlements.
>
> (Colorado River Delta Bi-National Symposium Proceedings 2001: 55).

It was also noted that

> Alternatives for providing the water needed to meet ecological needs should be determined. These could include recognizing the Delta as a

> legitimate user of Colorado River water; buying or giving water rights to the Delta; or using agricultural irrigation surplus from both sides of the border.
>
> (Colorado River Delta Bi-National Symposium Proceedings 2001: 56–7)

> The shrimping crisis at the end of the 1980s and into the beginning of the 1990s, resulted in a 50 percent reduction in shrimp catches. This crisis corresponded with the five consecutively driest years of Colorado flow to the Gulf. Likewise the collapse in the totoaba fishery in the early 1970s followed the opening of Glen Canyon Dam.
>
> (IBWC 2001: 57)

> Internationally, the vaquita was acknowledged in the 1980s and the importance of the Delta in the 1990s. The Mexican federal government, through the Secretariat of the Environment, began elaboration of the management program starting in 1996.
>
> (IBWC 2001: 59)

Colorado River delta

The Colorado River forms a delta where it flows into the Gulf of California, which once covered 8,612 km^2 and supported fertile plant, animal and aquatic life (Alles 2006). Sediment from the high desert areas was transported down the river to build up the delta until the damming of the river in the twentieth century trapped it in the reservoirs. With most of the water abstracted for irrigation and urban water supply, the means of sediment transport are no longer available. The delta is now largely dried out, with much of the former vegetation reduced and the formerly freshwater is brackish and saline (IBWC 2001), as mudflats; coastal wave action is now eroding the area. The delta area has been very considerably reduced and Alles cites the area as only 10 per cent of the former (Alles 2006). The delta area is protected under a range of designations, such as the UNESCO biosphere reserve since 1993 (IBWC 2001: 58), together with the upper Gulf of California area, and also a 2,500 km^2 Ramsar Wetland under the UN Convention on Wetlands for areas of international importance in terms of species, ecology and hydrology.

The amount of water estimated to sustain the delta would be approximately 436×10^6 m^3 over four years or 40×10^6 m^3 per year, and 316×10^6 m^3 every fourth year, or less than 0.5 per cent of the total run-off for the Colorado River over this time (IBWC 2001: 41).

The Colorado River's environmental characteristics can be summarised as:

- peakiness of flows
- variability of flow
- relatively low mean flow

- extreme dependence on river by many categories of 'environment'
- sensitivity of environment to changes to river regime
- multi-purpose use of water resource
- international nature of water issue
- large number of dams
- highly managed river.

A summary of impacts for the Colorado River is given in Table 5.6.

Table 5.6 Impacts linked to the major Colorado River HEP schemes

- Loss of seasonal flow variations.
- Attenuation of high and low flows.
- Pulse releases below dams, e.g. Glen Canyon (Petts 1984: 266).
- Trapping of sediment.
- Channel degradation, e.g. 2.6 m below Glen Canyon (Petts 1984: 121, 123–5, 130).
- Channel sedimentation below tributary confluences (Petts 1984: 133).
- Vegetation changes due to loss of flood flows and sedimentation (Petts 1984: 138).
- Loss of indigenous fish species below dams due to bottom discharge cold water releases (Petts 1984: 214, 216).
- Water quality changes.
- Change from warm turbid water quality to cold clear water, changes in fish habitat and range.
- Abstraction of water.
- Evaporation losses.
- Saline water quality in lower reaches and delta.
- Barriers to fish migration.
- Erosion of delta due to sediment transport loss.

5.2.4 Columbia River

The Columbia River rises in western Canada in British Columbia and flows between latitudes 55°–47°N for 1,931 km (Lang 2007). It springs from two lakes between the Continental Divide and the Selkirk mountain ranges at a height of 807 metres (Lang 2007). It flows north for over 321 km, before turning south to the border with the USA, then flowing south-west, skirting one of the Columbia plateau's extensive lava flows. The river turns south-east and flows through the volcanic shield in a gorge, towards the confluence with the main tributary, the Snake River. From here it flows westward to the Pacific.

The river catchment basin of 670,810 km^2, with its ten main tributaries, drains a variety of climatic and ecological areas, from temperate rainforests to semi-arid plateaus (Lang 2007). Precipitation ranges from 2.79 m to 0.1524 m. Melting snow contributes much of the discharge, and this fluctuates seasonally, with the highest volumes in April to September and the lowest in the winter months from December to February. The average river gradient is over 0.038 per cent, but in some sections is more than double this. There are several former falls and rapids sections, such as the Celilo Falls and Cascades Rapids, now inundated by reservoirs for the HEP schemes (O'Connor 2004).

The Columbia River, together with its main tributary, the Snake River, and smaller tributaries, drains an area the size of France and was once one of the most prolific fisheries of salmon and steelhead migratory fish – so much so that the wild salmon of the north-western area of the USA, covering several states, was central to native populations' economy, traditions and culture, as a totem. After the Second World War the Federal Government together with the individual states' administrations carried out a development programme for the area, with HEP development of the large rivers central to the plan. Other purposes of the dams were to raise water levels for navigation, so that ocean-going ships now reach Portland, and barges can reach the inland port of Boise in Idaho. Also, the reservoirs are used considerably for irrigation purposes and have contributed greatly to agriculture in the dry plateau areas. The cheap hydro electricity has aided industrial development to a large extent, e.g. aluminium smelting and the Hanford nuclear site are powered by it.

The Columbia River has been described as the most hydro electrically developed river system in the world (Lang 2007). The river's gradient profile has been turned into a series of steps, with only the Hanford reach and the tidal reach below Bonneville remaining free-flowing in the US section, unimpeded by impoundments. This can be seen from the gradient profile.

Today, the Columbia River has a series of 14 large dams in Washington State and in British Columbia, with a combined installed capacity of ~24–25 GW forming a significant part of the total HEP capacity of the USA (FCRPS 2003). There are 12 more dams on the large tributary, the Snake River, most of which have turbines for power generation.

There has been a considerable decline in the populations of migratory fish, especially in the period since the dams were constructed. Only ~2 per cent of the former numbers of fish now migrate up the runs (Joint Staff Report 2006). This has become a major environmental and political issue, since the fish are protected in a Treaty of 1855 between the USA and the Nez Perce Tribe, Confederated Tribes of the Umtilla Indian Reservation, Confederated Tribes of Warm Springs of Oregon, and the Confederated Tribes and Bands of the Yakama Nation (North West Council 2002) with the Native American Indian population. This issue is the focus of much work, study and expenditure, as well as considerable controversy.

Considerable efforts have been made to equip the dams for passing migratory fish. Fish ladders, turbine tube intake barriers, diversionary channels, trucking and barging the small fry and smolts around dams are some of the main methods. However, the success rate has not been high. The changes effected to the river, from a free-flowing state with a series of

Table 5.7 Impacts linked to the Columbia River HEP schemes

- Loss of salmon habitat.
- Disruption of salmon migration.
- Blockage of salmon migration routes.
- Fish barrier effects.
- Fish damage.
- Degradation of water quality.
- Evening out of seasonal flood flows.
- Flood control.
- Inundation of land.
- Displaced land use, e.g. farming.

fast-flowing shallow rapids with the occasional falls, and quieter pools, to a succession of wide, deep, slow-flowing lakes, confined by hazardous dam passages, has so altered water conditions for migratory fish as to 'wreck their original habitat' (Lang 2007).

Table 5.7 summarises the main impacts of the Columbia River HEP schemes.

5.3 Methods of analysis

This section describes the methods of analysis of data from the cases studied, as applied to the hypothesis. First, the hypothesis was tested against the parameters of the dam and reservoir, and then against the flow regime of the river below the dam for periods before and after construction of the dam, where data were available to perform this.

5.3.1 The HEP reservoir and dam parameters

Quantitative data for a list of environmentally significant HEP plants were collected in order to test for correlation with the proposed key parameters for environmental impact. A spreadsheet was constructed with key HEP data such as head height, mean flow rates and annular area of turbine tubes, to provide a figure for maximum power flux density. The aim was to discover whether any linkage could be found between the power flux density and the impacts.

As part of this, the aim was to compare the power flux density of the HEP plant with that of the river in an 'unchanged' state, i.e. its former power flux density. This would demonstrate the magnitude of the change in power flux density terms for the river over that part of the course now 'flattened out' by the backwater from the dam impoundment.

A set of data parameters for the HEP schemes was selected, based on head heights, rivers flows, and reservoir and dam characteristics. A list of the data parameters for the HEP schemes is shown in Table 5.8.

Table 5.8 List of data parameters used in test

Parameter	*Unit*	*Comments*
Reservoir area	km^2	
Basin volume	10^6 m^3	
Dam height	m	
Head height	m	
Turbine flow rate Q (max)	m^3 s^{-1}	
Flood river flow rate	m^3 s^{-1}	
Mean river flow	m^3 s^{-1}	
River minimum discharge	m^3 s^{-1}	
Compensation flow (mean)	m^3 s^{-1}	
Spillway	m	
Dam crest length	m	
Installed capacity	MW	
Turbine type		
Turbine diameter	m	
Number of tubes		
Annular area of tubes	m^2	
Annual average electricity production	GWh	Annual average electricity production
Energy	TJ	
Power flux (river) at dam line	kW	
Production factor	Ratio	Derived calculation
Capacity factor	Ratio	Derived calculation
Installed capacity per reservoir area	kW m^{-2}	Derived calculation
Power flux density	kW^{-2}	Derived calculation
Energy flux across converter	TJ m^{-3}	Derived calculation
Sediment trap efficiency of reservoir	Ratio	Derived calculation
Reservoir length	m	
Old river course mean gradient (reservoir)	Ratio (m/m)	Derived calculation
Mean water residence time	Years	Derived calculation
Sediment load	10^6 t per annum	
Mean reservoir cross-sectional area	m^2	Derived calculation
Mean reservoir fluid velocity	m s^{-1}	Derived calculation

The parameters for each hydro electric scheme were used to test for the proportion of energy extracted from the available energy in the flow, to discover how this varied between the different schemes, and whether it might be a significant factor in the impact experienced. Comparisons between the different schemes could be made on this basis.

The variable, proportion or ratio could then be tested for a link to impacts and the specific impacts experienced. In this study 'proportion' of the energy flow extracted and converted to electricity has been given the term 'production factor'.

5.3.2 Definitions of parameters

Basin volume

Values for basin volume have been obtained from data sources. However, the volume will depend on pool operating height for the reservoir. Ideally an average operating height and thus volume should be cited. A difficulty arises in comparing reservoir volumes with river basin volumes in 'run of the river' schemes, which leave the river bed relatively unchanged apart from some increased depth and width. The modified river basin may still exhibit some characteristics of river beds, though without very appreciable gradients, due to bed friction and available energy balances, as compared with the reservoir, which will be level.

Mean river flow $m^3 s^{-1}$

This has been taken wherever possible from flow at the dam line. In certain cases, *inflow* into the reservoir has been the measured parameter included (e.g. Glen Canyon). In other cases (e.g. Hoover dam, Lake Mead) water is abstracted from the reservoir for irrigation or water supply, and dam line flow figures can be misleading.

Flood river flow rate $m^3 s^{-1}$

This has been taken from the maximum recorded flood in the measurement period. Other possible representations are yearly maximum average, or ten-year maximum flow rate or 100-year maximum flow rate.

Mean former river gradient

The mean former river gradient has been derived by dividing the head height by the length of the reservoir. The reservoir surface is level, and since the method takes no account of the depth of the river at the reservoir inflow point, which may be substantial, or downstream of the dam, it is assumed that the gradient is that of the water surface. Using the mean gradient for the Belgrade–Iron Gates reach cited in the Joint Danube Survey (2005), of 0.04 per cent, and taking the head height of the scheme as 34 m, gives a level reservoir length of 85 km. Elsewhere, the length of the reservoir is cited as 112 km (Cioranescu 1980). This may be explained by varying water levels and the effects of the scheme extending further upstream than the reservoir itself. Where possible, data for river and reservoir elevations should be cited in preference to this approximation. Data for gradients on the Danube are taken from the Joint Danube Survey (Joint Danube Survey 2005: 23), and also GRDC data and elevation data obtained from published papers and from Google Earth satellite images.

For the Nile, data for the gradient slope profile was obtained from Rzoska (1978) and also Marndouh (1985) and Sutcliffe and Parks (1999). Gradients have been cross-checked wherever possible by using elevation data from sources such as Google Earth aerial and satellite images. The resolution level obtained is of necessity somewhat coarse, and subject to seasonal variations in the case of water levels.

Mean reservoir cross-sectional area

This has been derived by dividing the reservoir capacity at mean pool elevation, by the reservoir length, which is an approximation used by the US Army ACE (1989).

Mean reservoir fluid velocity

This has been derived by dividing the inflow rate by the cross-sectional area, a technique used to approximate reservoir fluid velocity by the US Army ACE (1989).

The data for the parameters for 11 cases of HEP schemes and a further six cases were assembled into a spreadsheet.

5.3.3 Details of analysis method

The land area of the HEP scheme has been obtained from the reservoir area. This can then be compared with the annual average generation output, to provide a watts per square metre value. The annual average generation can be divided by the installed capacity to provide a value for the capacity factor.

By using the equation:

$$P(in\,kW) = 10 \times Q \times h$$

Where:
P = power
10 = approximation to 9.81 m s^{-2} (acceleration due to gravity)
Q = $m^3\ s^{-1}$
h = head height

(Ramage 1996)

the theoretical average power in the HEP scheme can be calculated from the average flow rate, which can then be compared with the average annual output of the scheme, and allowing for an efficiency factor, this can be used to calculate the proportion of the energy extracted for conversion to electricity.

5.3.4 Possible impact parameters

- *Land use or land take*
    - m^2 per kW installed, i.e. power
    - m^2 per kWh produced, i.e. energy
    - km^2 per TWh p.a., i.e. average power
    - higher = more impact
- *Population displacement*
    - capita numbers
- *Flow changes*
    - river course to reservoir difference
- *Water speed changes*
    - river to reservoir changes upstream of dam m s^{-1}
    - reservoir to river changes downstream of dam m s^{-1}

Sedimentation

Sediment originates not just in the upper reaches of rivers, but all along the bed, and is generally in dynamic balance between processes of erosion, transport and deposition (Petts 1984). It has been estimated that reservoirs trap about 25–30 per cent of the global sediment being actively transported in rivers (Vorosmarty *et al.* 2003, cited in Owens *et al.* 2005). Suspended sediment transported to the global ocean has been estimated to be of the order of 15–20 $\times$ 10^9 t per year, with the larger proportion being discharges by rivers in mountainous areas (Milliman and Syvitski 1992; Farnsworth and Milliman 2003; Syvitski *et al.*, 2003, cited in Owens *et al.* 2005).

Sedimentation might be measured in rate of deposition tonnes per year, or in loss of deposition downstream tonnes per year, or as a percentage of the river's load. Alternatively, the reservoir trap efficiency might be measured (Toniolo *et al.* 2007). Reservoir sedimentation is dependent on three factors: size of the reservoirs drainage basin; characteristics of the basin which affect the sediment yield; and ratio of the reservoir's storage capacity to the river's flow (Petts 1984).

Sediment transport depends on water speed. Langford (1983) notes that below 0.2 m s^{-1} water movement, silt and mud will settle out, while at velocities of 0.2–0.4 m s^{-1} stream beds would comprise sand among the stones and gravel. Sediment transport increases as the water speed increases, and also larger diameter particles are transported. Settling velocity versus particle diameter equations of both linear and non-linear form have been proposed (Graf *et al.* 1966; cited in Graf 1971: fig. 4).

The efficiency of a reservoir as a trap for sediments has been described by the Brune Curve (ICOLD 1989). The Brune curve is an empirical tool describing how the longer the residence time in a reservoir (volume/inflow), the greater the sediment trap efficiency (Brune 1953). However, it does not describe the physical principles involved. Other methods of estimating trap efficiency have been developed, e.g. by Brown (1944) and Churchill (1948), as well as the US Army ACE (1989) and others. Although more sophisticated methods and models have been developed since, these methods were considered sufficient for a coarse analysis.

Brown's method of sediment efficiency calculation employs reservoir capacity and watershed area and is useful where the only parameters known are the storage capacity and the watershed area. Brune's method is considered more accurate than Brown's, but should only be applied to normal ponded reservoirs, not desilting basins or semi-dry reservoirs (US Army ACE 1989). Churchill's method uses a parameter, the sedimentation index, which is the period of retention divided by the reservoir mean velocity (Churchill 1948). The reservoir length needs to be known for this.

Brune's curves, empirically derived, assuming a median curve and some additional data, were described in this predictive equation by Dendy:

$$E = 100 \times (0.97^{0.19^{\log C/I}})$$

Where:
E = reservoir sediment trap efficiency
C = reservoir capacity
I = inflow rates per year.

By contrast Brown's Curve is described by the equation:

$$E = 100[1 - 1/91 + (KC/W)]$$

Where:
E = reservoir trap efficiency (percentage)
C = capacity (in acre-ft)
W = watershed area (in square miles)
K = coefficient ranging from 0.046 to 1.0, with median value 0.1. K increases for regions of smaller and varied retention time (calculated using the capacity–inflow ratio) (US Army ACE 1989).

Churchill's method (Churchill 1948)

Churchill's method of estimating trap efficiency uses a sedimentation index, which is the period of retention divided by the reservoir mean velocity. The retention time (R in seconds) can be estimated by the reservoir capacity divided by the daily inflow rate. The reservoir mean velocity is obtained from either field data or approximated from the average daily inflow rate divided by average cross-sectional area. The average cross-sectional area can be obtained by dividing the capacity by the reservoir length.

So:

$$S.I. = R/V$$
$$R = C/I$$
$$V = I/A$$
$$A = C/L$$
$$S.I. = CA/I^2 = (C/I^2)\ (C/L) = (C/I)^2/L$$

Where:

$S.I.$ = sedimentation index
R = retention time in seconds
V = mean velocity in ft per second
C = capacity in cubic ft
I = inflow rate in cubic ft per second
A = average cross-sectional area in square ft
L = reservoir length in ft at mean operating pool elevation

(US Army ACE 1989).

Following on from these impact parameters, caused by flow changes, are other effects which could be included as parameters:

- *Water quality reduction*
    - Temperature change °C
    - Chemical, e.g. total dissolved gases (TDG)
    - Oxygen dissolved, per cent
- *Nutrient loss*
    - Organic materials carbohydrate flow
- *Changed species*
    - Indigenous species change
- *Introduced species change*
    - Biodiversity
    - Percentage species reduction.

The barrier of the dam itself creates impacts; possible parameters are:

- *Fish barrier effects*
    - Percentage rate of passage as proportion of pre-dam passage
- *Fish damage*
    - Strikes by turbine: proportion
    - Pressure changes

However, regulating the flow has some positive effects for human use too:

- *Flood protection*
    - Incidence/severity of flooding
- *Flow regulation*
    - Ensuring even flows throughout the year
- *Navigation*
    - Increased water depth and draft.

5.3.5 Test for hypothesis parameters

The sample was tested for power flux density, using head height plus flow rate, for each HEP scheme. Hydrostatic pressure was also estimated. A test for the proportion of the energy flow extracted was carried out, where known compensation flows were taken into account. A test for the efficiency of energy conversion was made by comparing the production factor with the theoretical energy available from the average river flow. The number of conversions of energy form was determined.

These parameters were assessed for correlation and causal linkage with impacts. The parameters above were tested against the following measures of impact.

- area occupied (land use), i.e. reservoir area;
- lacustrine environment created, i.e. flow rate change/residence time;
- reduced load transport downstream, i.e. flow rate change (decrease);
- sedimentation, i.e. sensitivity to mechanical energy loss (dependant on sediment load and decrease in water current);
- flow changes downstream, i.e. losses, abstraction and flow fluctuation;
- increased erosion downstream, i.e. flow fluctuations.

Other parameters could be tested but were beyond the scope of this study:

- water quality changes, i.e. changes in temperature, chemical changes;
- eutrophication, i.e. water quality, oxygenation;
    - residence time;
- barrier to migratory fish;
    - fish ladder by-pass, head height;
    - upstream/downstream;
    - compensation flows;
- fish damage from pressure changes, i.e. pressure changes across turbine converter;
- turbine fish strikes, i.e. turbine design.

5.3.6 Testing the hypothesis against the river flow regime below the dam, before and after the HEP scheme.

Stream power

This concept described in Chapter 3 was originated by Bagnold (1966). He states that 'The available power supply, or time rate of energy supply, to unit length is clearly the time rate of liberation in kinetic form of the liquid's potential energy as it descends the gravity slope S.' Bagnold denotes this as:

$$\Omega = \rho g Q S \qquad (1)$$

where:
Ω = stream power ($W\ m^{-1}$)
ρ = density of water ($kg\ m^{-3}$)
g = acceleration due to gravity ($9.8\ m\ s^{-2}$)
Q = discharge ($m^3\ s^{-1}$)
S = slope (dimensionless).

Stream power represents the rate of energy dissipation through friction and work, with the bed and banks, per unit length of a stream. It is assumed that the water is not accelerating and the channel cross-section remains constant. These are reasonable assumptions for an averaged reach over a fairly short distance. If there is no acceleration of the water flow and the cross-section remains constant, it follows that the energy of the water flowing down a slope must be being dissipated in friction and in work done. The work done would be channel incision or sediment transport, while friction with the bed and banks would convert energy to heat.

Bagnold's concept can be employed in a number of different ways, and various forms of the general power equation have been developed. Stream power is used to study sediment transport, bed-load movement and to determine whether a stream is aggrading (accumulating sediment) or degrading (losing sediment or incising), as well as for channel pattern and bank-side habitat development.

Stream power can be used to calculate the power of the stream per unit length, as above or per unit bed area as below. Bagnold states that

> the mean available power supply to the column of fluid over unit bed area, to be denoted by ω is therefore:
>
> $$\omega = \frac{\Omega}{\text{flow width}} = \frac{\rho g Q S}{\text{flow width}} = \rho g d S u \qquad (2)$$
>
> Where:
> ω = available power per unit length and unit width
> ρ = density of water

d = channel depth
g = acceleration due to gravity
u = average fluid velocity
S = slope
Q = flow rate

(Bagnold 1966)

Bagnold equates the rate of doing work with the available power multiplied by the efficiency. The available power ω is then the supply of energy to overcome friction and for the transport of sediment.

With the increasing use and development of the stream power concept, the wide variety of different terms and expressions used became confusing, leading Rhoads (1987) to propose a standardised nomenclature. This nomenclature is employed in this book. Two specific uses of the stream power concept are employed in this study, (1) stream power per length of river reach, termed 'total stream power'; and (2) stream power per metre of river, termed 'cross-sectional stream power'.

Stream power, used to define the power of a defined reach of a stream channel, in watts, known as 'total stream power' or TSP (Rhoads 1987), is defined as:

$$p = \rho g Q S X \tag{3}$$

where:

p = total stream power in W
ρ = density of water in kg
g = acceleration due to gravity in m s^{-2}
Q = flow rate in m^3 s^{-1}
S = slope h/l in mm^{-1}
X = length of reach in m.

Stream power per unit length of a defined reach (W m^{-1}), termed cross-sectional stream power or CSP, is defined as:

$$\Omega = \gamma Q s \tag{4}$$

where:

Ω = cross-sectional stream power (W m^{-1})
γ = specific weight of water (9,810 N m^{-3})
Q = water discharge ($m^3 s^{-1}$)
s = the energy slope (m m^{-1}) which may be approximated by the slope of the channel bed.

(Rhoads 1987)

Using Equation (3) for total stream power, this was calculated for the sample rivers and hydro electric schemes over the extent of old river course now occupied by the reservoir, to give a value in kilowatts. The results are contained in the next chapter. This concept can be used for finding out the total power available over the average gradient of the former river bed, and could provide an indication of the amount of energy available for performing the river functions, e.g. sediment transport, albeit at a coarse resolution level. The cross-sectional stream power was then calculated for the old river course now occupied by the reservoir, using Equation (4).

Averaging out the former river bed gradient over the reservoir length will mask the many likely small-scale gradient variations, such as falls and rapids that are frequently drowned by the reservoirs of hydro electric schemes. However, the technique could give useful indications of the energy and functions balance of that section of the river bed now changed, in terms of inputs and outputs. As far as the author is aware, this use of total stream power concept and cross-sectional stream power in connection with environmental impacts from HEP schemes and reservoirs is novel.

5.3.7 Applying stream power concept to the data set

Methods of deriving approximations of unknown parameters of reservoir bed and flow are as follows. The known parameters are:

- Q = flow rate (cubic metres per second)
- S = slope gradient (fraction)
- specific weight of water = 9,810 N m^{-3}).

The unknown parameters are:

- B = channel width (metres)
- D = depth (metres)
- u = average fluid velocity (metres per second).

Methods of deriving approximations for unknown parameters are as follows. For D depth in reservoir, we can assume an even gradient along the former river bed as far as the reservoir river entrance.The gradient will be the head/reservoir length. The average depth at any given point in the former river bed will then be approximated to the proportion of the head height given by the distance to the reservoir entrance, to the reservoir length. However, this method assumes that the river bed had effectively no depth at the reservoir entrance, though this may be cancelled out by the fact that the head height does not incorporate the (downstream) river depth. Average approximate reservoir depth can be found by dividing the volume by the surface area.

For u, the average fluid velocity, the inflow rate divided by the cross-section square area of the reservoir bed section is required. The cross-section area is obtained from the width and depth. Where these are not known, the average cross-sectional area can be approximated by dividing the reservoir capacity by the reservoir length (US Army ACE 1989) at the average operating pool elevation. The average fluid velocity is the inflow rate divided by the average cross-sectional area. However, this method will only provide the average velocity for the whole reservoir, which may be hundreds of kilometres long and vary greatly in width and depth.

For B width, if this parameter is unknown, then the average width can be obtained in the following manner.

The equation

$$\omega = \frac{\rho g Q S}{\text{flow width}} = \rho g d S u \tag{2}$$

can be expressed as

$$\omega = \frac{\gamma Q S}{B} = \gamma D u$$

Where:
γ = specific weight of water (9810 N m^{-3})
B = breadth
D = depth
u = average fluid velocity.

Rearranging the equation,

$$\omega = \frac{\gamma Q S}{B} = \gamma D u$$

will provide

$$B = \frac{Q}{Du}$$

(note that there is effectively no slope), i.e.:

breadth = flow rate / depth × water velocity

However, this can only provide an average breadth for the reservoir, and only an approximation at that. Even so, this could be a useful guide to the values to be expected, in the absence of the actual data. Where possible, these approximate values should be validated with actual data.

5.3.8 Manning's equations

Another widely used equation in hydrology is that of Manning (1889), based on empirical observation and testing. Manning's equation is used to predict the behaviour of open-channel flows. This is a partly empirical equation used for modelling water flows which are open to the atmosphere and are not under pressure; though superseded by more sophisticated models, it is still employed today. Manning's equation for velocity is:

$$V = (1.49/n)\,R^{2/3}\,S^{1/2}$$

Where:
V = velocity (m/s)
n = channel roughness (value obtained from table)
R = hydraulic radius ($m^2\ m^{-1}$ (R = cross-sectional area/wetted perimeter)
S = slope ($m\ m^{-1}$).

R hydraulic radius is found by dividing the cross-sectional area of the water by the wetted perimeter.

Manning discovered by experiment that the flow velocity varies as a square root of the slope of the channel, the shape of the channel, i.e. its depth and width, and the roughness of the channel. The slope of the channel determines the gravitational force exerted per metre of bed. The shape of the channel, its depth and breadth, determine the degree of friction incurred by the flow. The roughness determines further the flow efficiency and the degree of laminar flow achieved.

5.3.9 Dimensional analysis

This test brings together different concepts concerning river power density, i.e. $W\ m^{-2}$, in the hypothesis with the $W\ m^{-1}$ of the stream power concept. It is therefore very important to check rigorously that consistent dimensions are being used in the parameters employed. This is done by dimensional analysis (Sleigh and Noakes 2009). The units employed have been analysed below in terms of the basic physical dimensions of which they are comprised, to ensure that like is being compared with like. This is particularly important as units and terms used in the literature tend to vary quite widely.

The prime units used here are those of flow rate Q, expressed as cubic metres (m^3) of water, per second.

$$Q = \frac{m^3}{s}[L^3]/[T]$$

Acceleration due to gravity is expressed as

$$g = m/s^2[L]/[T]^2$$

Velocity is expressed as

$$v = \frac{m}{s}[L]/[T]$$

Where:
L = length
T = time
M = mass

Power is expressed as

$$p = kW\,[M] \times [L]^2 \times [T]^{-3}$$

Density is expressed as

$$\rho = kg/m^3\,[M]/[L]^3$$

Slope is expressed as

$$S = m\ m^{-1}\ [L]/[L]$$

Length of reach is expressed as

$$X = m\,[L]$$

Therefore, for the TSP equation $p = \rho g Q S X$ (3)
(Rhoads 1987)

$$[M] \times [L]^2 \times [T]^{-3} = [M]\ [L]^{-3}\ [L]\ [T]^{-2}\ [L]^3\ [T]^{-1}\ [L]\ [L]^{-1}\ [L]$$

This balances the equation in dimensional terms, making it consistent.

5.3.10 Application of stream power to the Nile, Danube, Columbia and Colorado Rivers below major HEP schemes

The longitudinal gradient profiles for the seven rivers were plotted downstream of the HEP schemes in the sample, from a variety of sources (GRDC 2008; Joint Danube Survey 2005; Google Earth). This process involved a total of four spreadsheets in detail and three further ones.

Flow data were obtained from GRDC, downstream of Khartoum for the Nile, and downstream of Vienna for the Danube. These data, covering the period 1871–1984 for the Aswan dam, and 1901–97 for Bratislava on the Danube, were then reassembled in the form of monthly mean flow hydrographs. This could then be used to show the flow regimes both before the major dams were constructed and after, and the associated changes. This process was repeated for the Columbia River downstream of a point below the Canadian border to the sea, a distance of 1,449 km, with monthly hydrographs assembled including, where possible, the period before the major impoundment dams and the period after. The Colorado River was similarly treated, subject to availability of flow records for the relevant period. This process involved a total of 51 spreadsheets.

Where flow data were available, the gradient profile spreadsheets were then extended to monthly mean flows and a calculation of the monthly mean total stream power for that particular reach. This could then show the difference in total stream power available to the river before and after the dam construction, and hence the loss of power available to perform erosion and transport of sediment.

Spreadsheets were assembled to show the total stream power and cross-sectional stream power for the Nile, the Danube, Columbia and Colorado Rivers according to the theory above. Mean flow rates were obtained from 11 GRDC measuring stations (Dornblut 2000) corresponding to seven reaches of the Danube, as surveyed by the Joint Danube Survey (2005). The gradient profile of the Danube River was obtained from the same source and checked using elevation readings from Google Earth satellite photographs. The resolution level is of necessity somewhat coarse as the dates of the composite aerial and satellite photographs are not known, and water levels can vary by several metres depending on the season. However, it is considered that a broad agreement of the figures has been obtained to produce a fair representation of the gradient profile within ~2 m in most cases.

Hydrographs, showing mean monthly river flows over periods ranging from 100 years to 40 years were produced for the Danube. Between 480 and 1,200 data readings were used per flow monitoring station, which is considered to be a fair sample. The effect of the two dams on flow rates was then assessed by calculation of the maximum and minimum flows, as well as the standard deviation, before and after the dam and reservoir construction. Total stream power was calculated for each of the seven reaches of the Joint Danube Survey, and also cross-sectional stream power per metre, at a coarse resolution.

5.4 Conclusion

The conclusion reached was that a viable test to investigate in broad terms the environmental impacts of HEP could be devised. The setting up of the test consisted of the selection of a set of cases of well-known environmental impact from HEP schemes, the identification of a set of parameters by which to measure the features of the reservoir reach and dam, the identification of impact parameters and the plotting of gradient profiles downstream from the schemes for four of the rivers, with energy flows estimated where data were available. The case sample is of necessity not a statistically representative one, in that the sample is too small, at 11 main cases, with a further six included.

The parameters of impact have been used where quantitative data were available, though this was not always possible. Sediment load data were not available for all the cases, and water quality, biodiversity changes and fish migration barrier data have been outside the scope of the test. The study has concentrated on the main energy parameters of land area and power flux density, flow rate changes and, by extension, its effects on sediment transport.

Only four of the rivers – the Nile, Danube, Columbia and Colorado – had gradient profiles plotted due to the extent of the task and the availability of data. River gradients were only plotted below the HEP schemes in question, since this is considered the major area of impact and because the rivers involved are some of the world's longest.

A variety of tests were applied to the sample of selected HEP schemes and rivers, aiming to test the hypothesis that the power flux density is related to the environmental impact. In particular, parameters of the dams and associated reservoirs have been investigated, as well as the river gradient and flow regimes downstream from these installations. The tests aim to compare and contrast the energy parameters of the river before and after installation of HEP schemes with the incidence of environmental effects and impacts. The power flux density, the land area used, the proportion or ratio of energy extracted from the flow, as well as the number of energy conversions involved, are some of the primary parameters here, as applied to the reservoir reach and the dam. Flow rate changes are significant measures of power flux density changes and these have been identified for the reservoir reaches. Sedimentation and sediment trap efficiency has been another parameter used where practical. Further possible impact parameters were proposed, but proved to be beyond the scope and resources available in this study.

The total stream power concept has been applied using Equation (3) – TSP = $\rho gQSX$ – and Equation (4) – CSP = γQs (Rhoads 1987) – to reaches

Figure 5.1 Three Gorges Hydro Electric Dam. Wikimedia Commons, source file: Le Grand Portage Derivative work: Rehman

Figure 5.2 Loch Laggan Hydro Electric Dam Scotland UK. Copyright Alexander Clarke ©.

of the four rivers that have had gradient profiles plotted below the major dams in the case study, in order to identify changes in total stream power and cross-sectional stream power from periods before dam construction to afterwards. Flow data from 30 river flow stations were collected, covering periods of up to 130 years in the case of the Danube. As stated earlier, to the author's knowledge the use of the concept of total stream power and cross-sectional stream power for determining impacts from HEP as a result of changes in power flux densities is novel. The complex and interrelated nature of these parameters has necessitated some considerable amount of detail and description here. The results are presented in the next chapter, showing the uses and application of the hypothesis in practice.

Bibliography

Alles, D.L. (2006) 'The Delta of the Colorado River', Western Washington University, online resources: www.biol.wwu.edu/trent/alles/TheDelta.pdf (accessed 26 February 2009).

Amin, K. (1994) 'Safety Considerations, Operation and Maintenance of Existing Structures of High Aswan Dam', *Water Power & Dam Construction*, January.

Bagnold, R.A. (1966) 'An Approach to the Sediment Transport Problem from General Physics', USGS, Professional Paper.

Borsos, B. (1991) 'Socio-Political Aspects of the Bos–Nagymaros Barrage System', *Water Power & Dam Construction*, May: 57–9.

Brown, C.B. (1944) 'Discussion of Sediment in Reservoirs by Witzig', *ASCE American Society of Civil Engineers, Transactions*, Vol. 109.

Brune, G.H. (1953) 'Trap Efficiency of Reservoirs', *American Geophysical Union Transactions*, Vol. 34 No. 3: 407–18.

Churchill, M.A. (1948) 'Discussion and Analysis and Use of Reservoir Sedimentation Data by LC Gottschalk', Federal Interagency Sedimentation Conference, Denver Colorado.

Cioranescu, G. (1980) 'Harnessing the Romanian Section of the Danube', Open Society Archives, online resources: http://files.osa.ceu.hu/holdings/300/8/3/text/53-2-83.shtml, accessed 22 July 2009.

Clarke, A. (1995) 'Environmental Impacts of Renewable Energy: A Literature Review,' Thesis for a Bachelor of Philosophy Degree, Technology Policy Group, Open University, Milton Keynes.

Colorado River Delta Bi-National Symposium Proceedings (2001) 'United States–Mexico Colorado River Delta Symposium 11–12 September 2001', online resources: www.ibwc.state.gov?FAO/CRDS0901/EnglishSymposium.pdf, accessed December 2008.

Dendy, F.E. (1974) 'Sediment Trap Efficiency of Small Reservoirs', *Transactions of ASAE*, Vol. 17 No. 5: 898–988.

Dickinson, H. and Wedgwood, K.F. (1982) 'The Nile Waters: Sudan's Critical Resource', *Water Power and Dam Construction*, January.

Dornblut, I. (2000) 'GIS Related Monthly Balance of Water Availability and Water Demand in River Basins. Case Study for the River Danube', Global Runoff Data Centre, Report No. 25.

Dubowski, Y. (1997) 'Environmental Effects of the High Dam at Aswan', June Talk at the California Institute of Technology, online resources: www.gps.caltech.edu/classes/ge148/1997C/Reports/148niled.html, accessed 23 July 2009.

El-Moattassem, M. (1994) 'Field Studies and Analysis of the High Aswan Dam Reservoir', *Water Power and Dam Construction*, January.

Encyclopaedia Britannica (2009) 'Colorado River', online resources: www.britannica.com/EBchecked/topic/126494/Colorado-River, accessed 13 August 2009.

Farnsworth, K.L. and Milliman, J.D. (2003) 'Long-Term Fluvial Sediment Delivery to the Ocean: Effect of Climatic and Anthropogenic Change'. *Global Planetary Change*, Vol. 39 No. 1–2: 53–64.

FCRPS (2003) 'Federal Columbia River Power System Brochure Guide', US ACE, Bonneville Power Administration, Bureau of Reclamation, Agencies, online resources: www.nwd-wc.usace.army.mil/PB/fcrps.broch.f.pdf, accessed January 2009.

Flessa, K.W., *et al.* (2001) 'Since the Dams: Historical Ecology of the Colorado Delta', Centro de Estudios de Almojas Muerlas, online resources: www.geo.arizona.edu/ceam/Hecold/hecolcd.htm, accessed 26 February 2009.

Graf, W.H. (1971) *Hydraulics of Sediment Transport*, McGraw-Hill, New York.

GRDC (2008) Data from the Global Runoff Data Centre, Koblenz, Germany.

Humborg, C. (1997) 'Effect of Danube River Dam on Black Sea Biogeochemistry and Ecosystem Structure'. *Nature*, Vol. 386: 385–8.

ICOLD (1989) 'Sedimentation Control of Reservoir Guidelines', International Commission on Large Dams.

IEA (2014) 'Key World Energy Statistics', International Energy Agency, online resources: www.iea.org/publications/freepublications/publication/KeyWorld2014.pdf, accessed 20 May 2015.

International Water Power and Dam Construction (1996) *Year Book*, Reed Business Publishing, London.

IBWC (2001) 'United States–Mexico Colorado River Delta Symposium 11–12 September 2001', online resources: www.ibwc.state.gov?FAO/CRDS0901/English Symposium.pdf, accessed December 2008.

Joint Danube Survey (2005) Online resources: www.icpdr.org/icpdr-pages/reports.htm, last accessed 22 July 2009.

Joint Staff Report (2006) 'Joint Staff Report Concerning Commercial Seasons for Spring Chinook, Steelhead, Sturgeon, Shad, Smelt and Other Species and Miscellaneous Regulations for 2006', Joint Columbia River Management Staff, Oregon Department of Fish and Wildlife, Washington Department of Fish and Wildlife, online resources: http://wdfw.wa.gov/fish/crc/jan1806jointstaff.pdf, last accessed 13 August 2009.

Kammerer, J.C. (1990) 'Largest Rivers in the United States', USGS, online resources: http://pubs.usgs.gov/of/1987/ofr87-242, last accessed 13 August 2009.

Lang, W. (2007) 'Columbia River History', Portland State University, online resources: www.ccrh.org/river/history.htm, accessed 18 October 2007.

Langford, T.E. (1983) *Use of Water for any Electricity Production, Hydro, Tidal, Cooling, Electricity Generation and the Ecology of Natural Waters*, Liverpool, Liverpool University Press.

Laszloffy, W. (1965) 'Die Hydrographie der Donau. Der Fluss als Lebensraum', in: Liepolt, R., ed., *Limnologie der Donau – Eine monographische Darstellung. Kapitel II*, Schweizerbart, Stuttgart.

Liska, M.B. (1993) 'Gabcikovo–Nagymaros: A Review of its Significance and Impacts', *Water Power and Dam Construction*, July.

Manning, R. (1889) 'On the Flow of Open Water in Channels and Pipes'. *Transactions of the Institution of Civil Engineers of Ireland*, Vol. 20: 161–6.

Marndouh, S. (1985) *Hydrology of the Nile Basin*, Elsevier, London, Amsterdam and New York.

Milliman, J.D. and Syvitski, J.P.M. (1992) 'Geomorphic/Tectonic Control of Sediment Discharge to the Ocean: The Importance of Small Mountainous Rivers'. *Journal of Geology*, Vol. 100: 525–44.

North West Council (2002) 'Native American Legal Resources', NW Council, online resources: www.nwcouncil.org/history/IndianTreaties.asp, accessed 15 April 2009.

O'Connor, J. (2004) 'The Evolving Landscape of the Columbia River Gorge Lewis and Clark and Cataclysms on the Columbia', *Oregon Historical Quarterly*, online resources: www.historycooperative.org/journals/ohq/105.3/oconnor.html, accessed January 2009.

Owens, P.N., Batalla, R.J., Collins, A.J., *et al.* (2005) 'Fine-Grained Sediment in River Systems: Environmental Significance and Management Issues', *River Research Applications*, Vol. 21: 693–717.

Petts, G.E. (1984) *Ecology of Impounded Rivers: Perspectives for Ecological Management*, John Wiley & Sons, Chichester.

Ramage, J. (1996) 'Hydroelectricity', in Boyle, G., ed., *Renewable Energy Power for a Sustainable Future*, Oxford University Press in conjunction with the Open University, Milton Keynes.

Rhoads, B.L. (1987) 'Streampower Terminology', *The Professional Geographer*, Vol. 39 No. 2: 189.

Rzoska, J. (1978) *On the Nature of Rivers*, Dr Junk Publishers, The Hague, Boston, MA, London.

Schwarz, U. (2008) 'Assessment of the Balance and Management of Sediments of the Danube', FLUVIUS Floodplain Ecology and River Basin Management, online resources: http://ksh.fgg.uni-lj.si/bled2008/cd_2008/05_Floods,%20morphological%20processes,%20erosion,%20sediment%20transport%20and%20sedimentation/211_Schwarz.pdf, last accessed 23 July 2009.

Sleigh, A. and Noakes, C. (2009) 'Dimensional Analysis, Notes for Level 1 Lecture Course in Fluid Mechanics', University of Leeds, School of Civil Engineering, online resources: www.efm.leeds.ac.uk/CIVE/CIVE1400/Section5/dimensional_analysis.htm, last accessed 22 August 2009.

Stankovic, S. (1960) 'Yugoslav–Rumanian Iron Gates Project', Open Society Archives Box Folder Report 122-1-230, online resources: www.osa.ceu.hu/files/holdings/300/8/3/text/122-1-230.shtml, accessed 29 October 2007.

Sutcliffe, J.V. and Parks, Y.P. (1999) *The Hydrology of the Nile*, International Association of Hydrological Sciences, Wallingford.

Syvitski, J.P.M., *et al.* (2003) 'Predicting the Terrestrial Flux of Sediment to the Global Ocean: A Planetary Perspective', *Sedimentary Geology*, Vol. 162: 5–24.

Toniolo, H., Parker, G. and Voller, V. (2007) 'Role of Ponded Turbidity Currents in Reservoir Trap Efficiency', *American Society of Civil Engineers, Journal of Hydraulic Engineering*, Vol. 133 No. 6: 579–95.

UNH/GRDC (2007) 'Danube Basin', online resources: www.grdc.sr.unh.edu/html/Polygons/P6142200.html, accessed various occasions.

US Army ACE (1989) 'Trap Efficiency of Reservoirs', US ACE United States Army Corps of Engineers, online resources: www.uace.army.mil/usace-docs/eng-manuals/em1110-2-4000/a-f.pdf, accessed 10 July 2006.

USGS (2004) 'Climatic Fluctuations, Drought and Flow of the Colorado River', Fact Sheet, USGS, online resources: http://pubs.usgs.gov/fs/2004/3062, last accessed 13 August 2009.

Vorosmarty, C.J., *et al.* (2003) 'Anthropogenic Sediment Retention: Major Global Impact From Registered River Impoundments'. *Global and Planetary Change* Vol. 39: 169–90.

WWF (2008) 'Danube Delta Fact sheet', World Wide Fund for Nature, online resources: http://assets.panda.org/downloads/danube_delta_factsheet_july2008_en.pdf, accessed 13 August 2009.

Zinke, A. (2004) 'Hydropower Station Gabcikovo: Deficits in Hydrology (Sediment Transport, Groundwater) and Biology', Int IAD Workshop 14–16 October, online resources: www.zinke.at/Zinke.data/Images/Zinke-IAD.Gabcikvo.122-04.pdf.

6 Test results

6.1 Conclusion to initial results

The hypothesis was tested using the case of hydro electric power (HEP) for a selection of HEP plants. There were two parts to the test; part one tested the parameters against data for the reservoir and dam, and the second part tested the hypothesis through the concept of stream power for the river reaches below the dams.

The results for part one of the test are summarised in Tables 6.1 and 6.2. It had been the intent to collect data for more parameters, as described in Chapter 5, but the main parameters for some of the main impacts of significance to this study have been included, apart from water quality data, which could be related to residence time, and other factors.

In part two of the test total stream power (TSP) calculations were estimated for four rivers in the sample, downstream of HEP schemes. Flow rates and TSP were compared for periods prior to the dam construction and post-dam construction. Average monthly flow rate data were collected and tabulated, and then monthly hydrographs drawn up. These were then incorporated into the gradient profile spreadsheets so that the TSP could be estimated. A summary of the main results is shown in Table 6.3.

There were discontinuities in both the timeline and the river line data sets as a result of the availability of flow data. However, despite these discontinuities and comparisons of flow over different time periods, a fairly representative version was obtained at a coarse resolution scale. Overall changes in climate and weather patterns are assumed not to be significant over these periods, and the main changes apparent are due to the operation of the HEP schemes reservoirs and water abstraction.

The Nile and the Danube had the longest continuous flow rate data records, in some instances from 1880 to the present, depending on the flow stations, though the Columbia and Colorado also had long-term flow records. The most geographically continuous set of flow station records was that of the Danube. In the case of the Nile, the majority of the flow stations below Aswan are fairly recent and cover only a limited period. However, a good to fair set of data was also obtained for the Columbia River and some

Table 6.1 Summary of reservoir and dam parameter data results

HEP Scheme	*Hypothetical power flux density ($kW\ m^{-2}$)*	*Mean river flow ($m^3\ s^{-1}$)*	*Production factor (ratio)*	*Land use ($km^2\ TWh^{-1}$ p.a.)*	*Former mean river gradient (%)*	*Average water residence time (days)*
Aswan High Dam	687	2,760	0.48	500	0.014	708.1 (1.94 years)
Kariba	932	1,775	0.55	687.5	0.038	1,177.49 (3.226 years)
Cabora Bassa	1,182	2,457	0.51	292.3	0.0446	296.77
Itaipu	1,162	9,000	0.82	17.9	0.0696	37.23
Hoover	1,555	395.6	0.75	159.75	0.13	1,024 (2.8 years)
Glen Canyon	1,525	875	0.30	185.6	0.0519	357.14
Bonneville	150	5,660	0.67	18.55	0.0211	1.35
Grand Coulee	987	3,100	0.78	15.6	0.129	44.02
Gabcikovo	211	2,047	0.79	20	0.0827	1.37
Iron Gates I	255	5,500	0.85	21.33	0.0232	5.26
Three Gorges	1,001	14,300	0.68	12.8	0.017	31.81

Table 6.2 Summary of reservoir and dam parameter data results

HEP scheme	*Reservoir volume (106 m^3)*	*Average reservoir fluid velocity ($m\ s^{-1}$)*	*Total stream power (kW)*	*Cross-sectional stream power ($W\ m^{-1}$)*	*Brune sediment trap efficiency (%)*
Aswan High	168,900	0.0082	1,896,000	3,790	98.13
Kariba	180,600	0.00246	1,654,000	6,620	98.7
Cabora Bassa	63,000	0.011	2,904,000	10,800	85.4
Itaipu	29,000	0.0528	10,454,000	61,500	96.53
Hoover	34,850	0.0014	612,000	4,970	98.56
Glen Canyon	27,000	0.0097	1,334,000	4,460	96.95
Bonneville	662	0.616	846,000		17.8
Grand Coulee	11,790	0.0332	1,846,000	14,600	86.93
Gabcikovo	243	0.219	432,000	16,600	18.14
Iron Gates I	2,500	0.594	1,403,000	12,500	52.3
Three Gorges	39,300	0.218	14,309,000	23,800	83.78

for the Colorado River. At Aswan (below the Aswan High dam) on the Nile, flow rate data from 1870–1984 was available, which covers the pre-dam period before 1964–7 (dam-filling period), and post-dam, for a more limited duration. At flow stations below Aswan, such as Gaafra, Esna, El-Hammadi and Nag Hammadi, only post-dam flow data were available for a limited period, 1973–84. Nevertheless, despite these data deficiencies, the Aswan flow data are considered by the author to be revealing, and its flow patterns are likely to be repeated at the lower flow stations. The very considerable loss of monthly peak flow TSP of 71 per cent post-dam operation is evident from the data, attenuating the August–September floods of the Nile's seasonal flow variations.

For the Danube, flow data were available for selected measuring stations from 1931 to 1999 and 2001 to 2002, covering the periods before the two dams' construction, the Iron Gates in 1970 and the Gabcikovo in 1992. Zimnicea flow station data, at 83 km below the Iron Gates scheme, covers the period 1931–2002, providing almost 40 years' data pre-dam construction and 30 since. At Zimnicea only a slight attenuation of peak flows can be discerned post-1970 and this appears to be within the normal bounds of variability. At Lom and Novo Selo, more limited flow data periods were available, 1941–70, and then 1991–99 for Lom and 1937–71 and 1992–99 for Novo Selo. TSP for the peak months of April was reduced by 6.9 per cent and by 10.4 per cent for May for the Novo Selo to Lom reach of 128.5 km. At Ruse, almost 400 km below the Iron Gates dam, a reduction of 10 per cent in TSP was shown post-dam. This reduction is likely to be within or near normal variation and so is not considered significant.

Table 6.3 Summary of TSP estimations downstream of HEP schemes before and after dams, for four rivers

River and HEP scheme	*Reach*	*Monthly peak total stream power pre-dam (kW)*	*Period*	*Monthly peak TSP post-dam (kW)*	*Period*	*Peak TSP Δ difference pre- and post-dam % (ratio)*
Nile Aswan High	Aswan–Gaafra	729,862	1871–1967	214,338	1970–84	–71%
Danube Gabcikovo	Dunaalmas–Nagymaros	86,789.07	1948–92	90,055.8	1992–95	+3%
Danube Iron Gates I	Novo Selo–Lom	244,193.9	1931–70	227,442.7	1992–99	–6.9%
Danube Iron Gates I	Zimnicea–Jantra	84,620.5	1931–70	81,775.1	1970–2002	–3.36%
Columbia Grand Coulee	Grand Coulee–Chief Joseph Dam	8,420,658.6	1930–73	3,740,111.5	1973–2005	–55%
Columbia	Below Priest Rapids–McNary Dam	2,050,525.4	1917–72	1,013,529.9	1973–2006	–50.57%
Colorado Glen Canyon	Lees Ferry–Little Colorado confluence	1,829,859.3	1911–63	527,974.2	1963–2003	–71.14%
Colorado Hoover	Yuma–Morelinos	82,003.7	1904–33	1,012.4	1960–75	–98.6%

At Dunaalmas, 82 km below the Gabcikovo HEP scheme, there is again very little difference in the pre- and post-dam flow rate regime (a 3 per cent increase in TSP), although the data period is too short to be representative as it extends to only three years post-dam (1948–95). Any diminution in monthly flow rate variability is likely to be the cumulative result of dams upstream (Schwarz 2008).

The Columbia River flow data covered the period 1930–2005, giving good pre- and post-dam construction periods (the 1941 Grand Coulee dam and 1973 Mica dam). The Grand Coulee dam is at the head of the US river reach estimated for TSP, and clearly shows the reduction in flow rate variability post-1973, although not very much in the period 1941–73.

Average monthly peak flows in June had diminished by 55 per cent, and there are now two peaks, one in June and a smaller one in January–February, which was formerly the lowest flow period. The reduction in flow rate variability continues downriver, as seen in the monthly hydrographs for Rock Island pool (reservoir reach), mean monthly peak flow reduced by 51 per cent. Below Priest Rapids dam, a 50 per cent reduction in peak flow and therefore TSP is seen, and the Dalles has a 45 per cent decrease in peak flows and TSP, while at Beaver Army Terminal the decrease in peak flows and TSP is 23 per cent. Data for Rock Island was available 1960–2005; for below Priest Rapids, from 1917 to 2006; for the Dalles 1878–2006; and for Beaver Army Terminal 1968–70. The significant drop in peak flows and TSP is considered to be the result of attenuation of flood flows by the Mica and other Canadian dams primarily, which regulate the flow.

The Colorado River, with two major impoundment dams, the Glen Canyon and the Hoover dam, exhibits great diminution of seasonal flow variation, since the dams have been designed for about one year's and 2.8 years' flow storage, respectively. A large proportion of the water of this river is abstracted for urban water supply and irrigation. A section of the river, the reach from the Mexican international boundary to kilometre 1,431 at the top of the Lake Powell reservoir was plotted for gradient and TSP at a coarse resolution. Data were available from 1911 to 2003 for Lees Ferry flow station 24 km below Glen Canyon dam, and this showed an average reduction in peak flow rates for June of 75 per cent after the dam's construction in 1963. The only other data point for which consistent data were available was Yuma, at 47 km above the international boundary for the period 1904–75. Data were also available for six months of 1983. The Yuma data clearly show the great changes in flow rates after the 1933 Hoover dam construction and the Glen Canyon dam in 1959. The data are thus described in three periods: 1904–33 pre-Hoover; 1933–59 pre–Glen Canyon; and 1959–75 post-dams.

The effects of attenuation of flood flows and associated water abstraction are clear; diminution of the average peak month flow rate for 1904–33, by 87 per cent, has occurred for the period 1933–59, with the peak now moved from June to January. A 98.6 per cent decrease in peak monthly TSP occurs

Table 6.4 A summary of some of the main impacts from the sample HEP schemes

HEP Scheme	*River*	*Impact*
Aswan High	Nile	Silting, loss of silt fertiliser downstream; erosion of delta, fishery loss, water loss evaporation from reservoir surface, salinity at delta, land use, population displacement
Kariba	Zambesi	Population displacement, flow fluctuation, water evaporation loss, wildlife effects, water-borne disease
Cabora Bassa	Zambesi	Flow fluctuation, downstream wetlands loss and silt loss, downstream terrain drying out, population displacement
Itaipu	Parana	Drowned vegetation – CO_2 emissions
Hoover Dam	Colorado	Flow fluctuations, water abstraction, flood attenuation, silting
Glen Canyon	Colorado	Water abstraction, flow fluctuations, silting
Bonneville Dam	Columbia	Migratory fish barrier, upstream river changes
Grand Coulee	Columbia	Supersaturation/(TDG) anadromous fish barrier
Gabickovo	Danube	Land drainage, groundwater changes, channel changes, ecosystem damage
Iron Gates	Danube	Sediment trap, land inundation
Three Gorges	Yangtze (Changjiang)	Population displacement, loss of habitat, sedimentation fears/forecasts, flow changes, silting

for the period 1960–75. The effects of water abstraction and losses due to evaporation can be seen in the average monthly flow rate reducing by 69 per cent, from the period 1904–33 to the period 1933–59, and then by about 97 per cent for the period 1959–75, representing losses therefore also of TSP.

Although only two flow station data points are included here for this river, they can be considered representative of the great changes in the flow regime, with consequent loss of TSP. The peak monthly TSP changes, i.e. losses, as shown in Table 6.3, to the rivers below the major HEP schemes could be considered to be the most significant impact of all the parameters used in this study.

A summary of some of the main impacts from the sample of HEP schemes is shown in Table 6.4.

6.2 Conclusions for the first part of the test on the reservoir and dam reach and the second part for the river reaches below the dams

It would appear that a relationship between the power flux density and some environmental impact parameters, but not for others, is suggested here for the sample HEP schemes chosen. However, due to the small sample, the results are not statistically significant.

The data suggested that land use per TWh p.a. may have an inverse linear correlation to power flux density in terms of cross-sectional stream power (CSP), but was not related to TSP for the reservoir reach. There was no evidence of a relationship between power flux density in terms of head height and land use for the sample. Water residence time suggested a linear relationship with land use, though it was not statistically significant. Land use was not likely to be influenced by the former average bed gradient in this sample. Reservoir volume, however, was likely to have a linear correlation to land use, which implies that variations in shape and depth can be discounted for the sample. There was no evidence for a relationship between average fluid velocity and land use. Some correlation between the production factor and land use was indicated, though only in broad terms distinguishing between water storage schemes and primarily generation schemes. There was no indication that mean river flow was related to land use.

Land area for HEP schemes depends largely on the reservoir size and is much less for the run-of-the-river schemes such as Bonneville and Gabcikovo, and also where flow rates are high, e.g. the Three Gorges, or where gradients and valley sides are steeper. Where there are high head heights but without large storage reservoirs, high power flux densities could result in low land use, but such examples were not included in this study.

The reduction in sediment transport, as reflected by loss of delta land area, indicated a linear correlation to reduction in TSP for the river downstream of the HEP scheme when tested for three rivers. TSP can be equated to power flux density and so this part of the test result is considered to be indicative of support for the hypothesis, though given the small sample cannot be stated with confidence.

Large-volume reservoirs relative to flows, with long residence times and big reductions in flow rates, were better sediment traps. In addition, such reservoirs, e.g. the Aswan High dam reservoir, the Hoover and Glen Canyon, attenuate flow variations more. As a consequence of this, downstream of the dam the changes, i.e. reduction of the power flux density as represented by the loss of the flood flow peaks, can result in reduction or near cessation, in the worst case, of one of the functions of the river flow, the transport of sediment. This has serious impacts on the recharging of deltas on the Nile and Colorado, which are both losing area as a result of coastal erosion. In this case it is the loss of power – that is, the rate of work – as well as the straightforward extraction of energy from the flow that is causing environmental impact. Extraction of the rivers' energy for electricity generation (or for water abstraction) does also result in losses to the functions, which could result in impacts. The greater the change in power flux density the greater the impact appears to be. This also applies to the reservoir reach, where the conversion of the kinetic energy transporting sediment to potential energy with a negligible current speed causes deposition of sediment. The larger the reservoir and the lower the current, the better a trap the reservoir is. The case of the Gabcikovo scheme – where 85 per cent of the flow is diverted away from the natural bed into a sealed artificial canal, causing drying of wetlands – demonstrates how the parameter proportion of interception and abstraction of the flow could be significant.

6.2.1 TSP for the four sample rivers Nile, Danube, Columbia and Colorado

River gradient profiles

In order to calculate the TSP for four of the selection of rivers, the longitudinal river gradient profiles were plotted using data from a variety of sources (GRDC 2008; Rzoska 1978; Marndouh 1985; Google Earth 2008 elevation), so that the stream power before and after dam construction could be calculated. In the case of the Nile between Khartoum and Cairo, the mean gradient is 0.0126 per cent, as the Nile falls from 375 m to 12 m over 2,873 km. However, this hides variations in gradient of 0.098 per cent at the Cataracts IV, to 0.0042 per cent below the Cataracts III. The zero gradient of Lake Nasser, the Aswan dam reservoir, can be observed from Figure 6.1. This reservoir has inundated the Cataracts I and II and therefore masks further gradient variations. The decline in gradient after Cairo, which comprises the delta region, can be discerned.

FORMER NILE RIVER GRADIENT PROFILE

The approximate former river gradient was plotted for the rivers for before construction of the Aswan dam and the High Aswan dam, as well as the impoundments lower down the river, for the section below Khartoum. This is shown in Figure 6.2. This shows the change in the profile, in broad terms, though no detailed profile data for the series of rapids now inundated by Lake Nubia and Lake Nasser were found. The same exercise was carried out for the Danube, Columbia, and Colorado Rivers.

DANUBE GRADIENT PROFILE

The mean gradient of the Danube below Bratislava is 0.00736 per cent, as the Danube descends from 139 m to sea level over 1,888 km. However, the gradient profile displays a gradual lessening in slope to the delta region. The two hydro electric scheme dams below Bratislava are both in relatively steeper reaches. The Gabcikovo scheme is at the end of the steeper Austrian–Slovakian section, with a gradient of 0.043 per cent, where the river flattens out in the Hungarian plain section to a gradient of 0.017–0.007 per cent and then to 0.004 per cent (JDS 2005). The Iron Gates section is steeper where the river cuts through the Carpathian Mountains in a narrow gorge, with a gradient of 0.027 per cent, as shown in Figure 6.3.

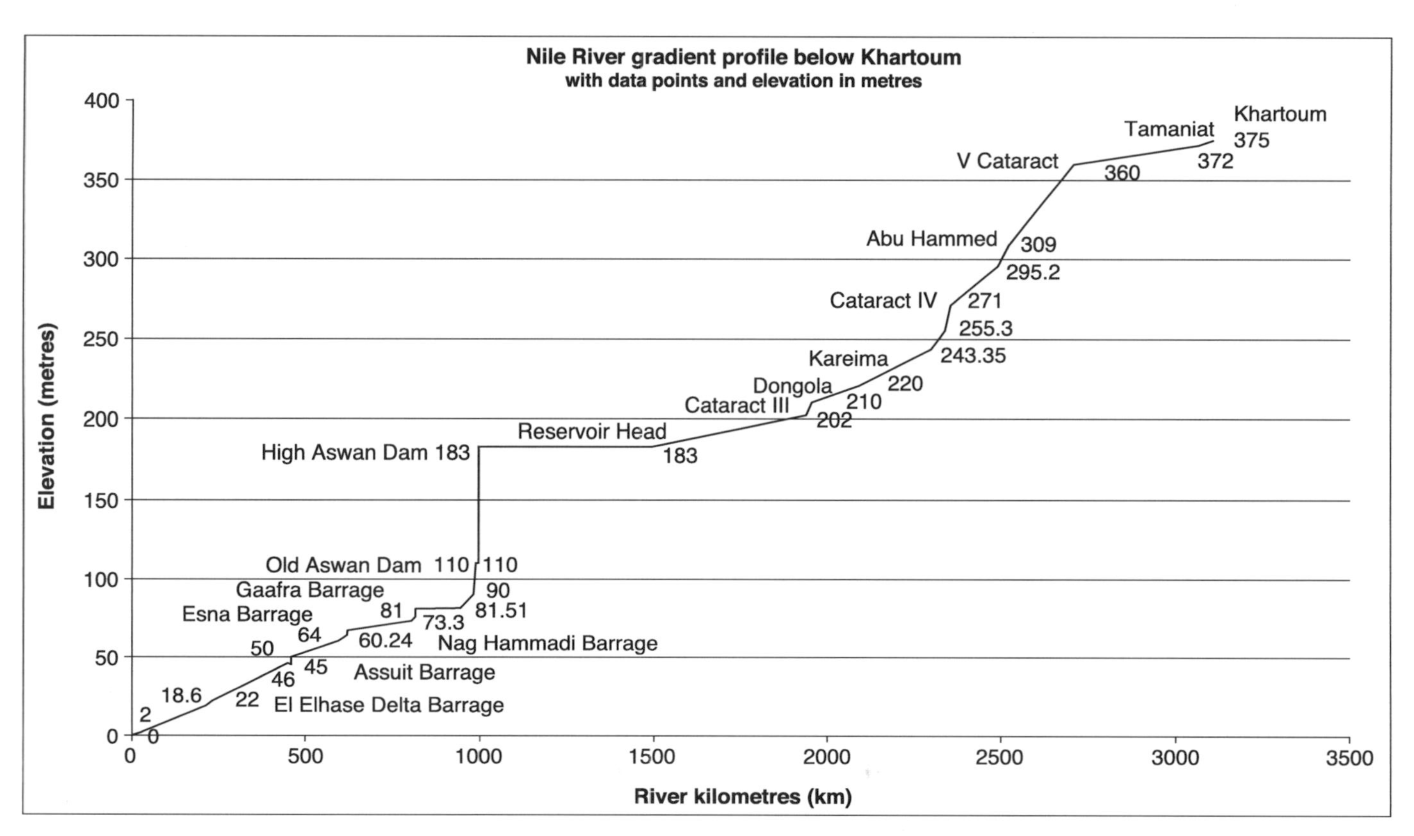

Figure 6.1 Nile River surface gradient profile below Khartoum before construction of the Merowe dam.

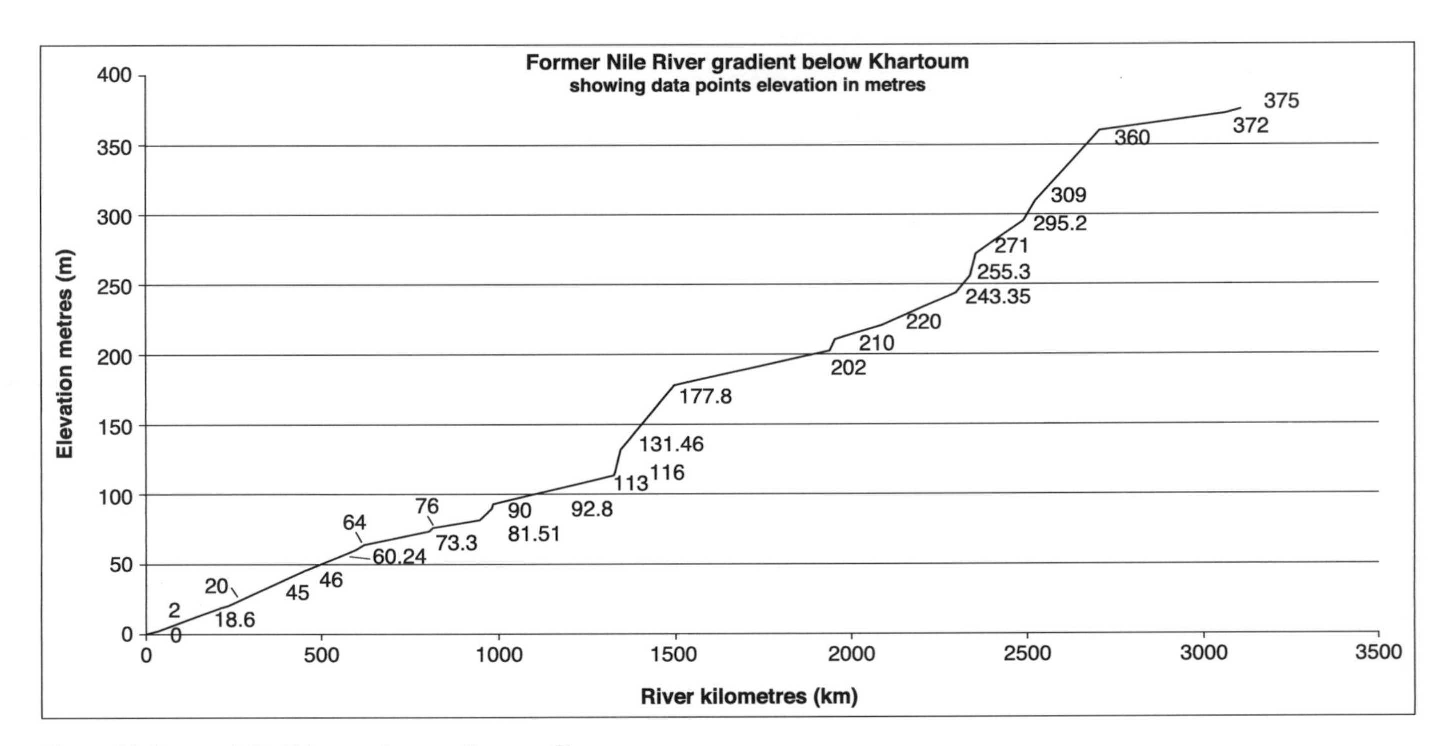

Figure 6.2 Former Nile River surface gradient profile.

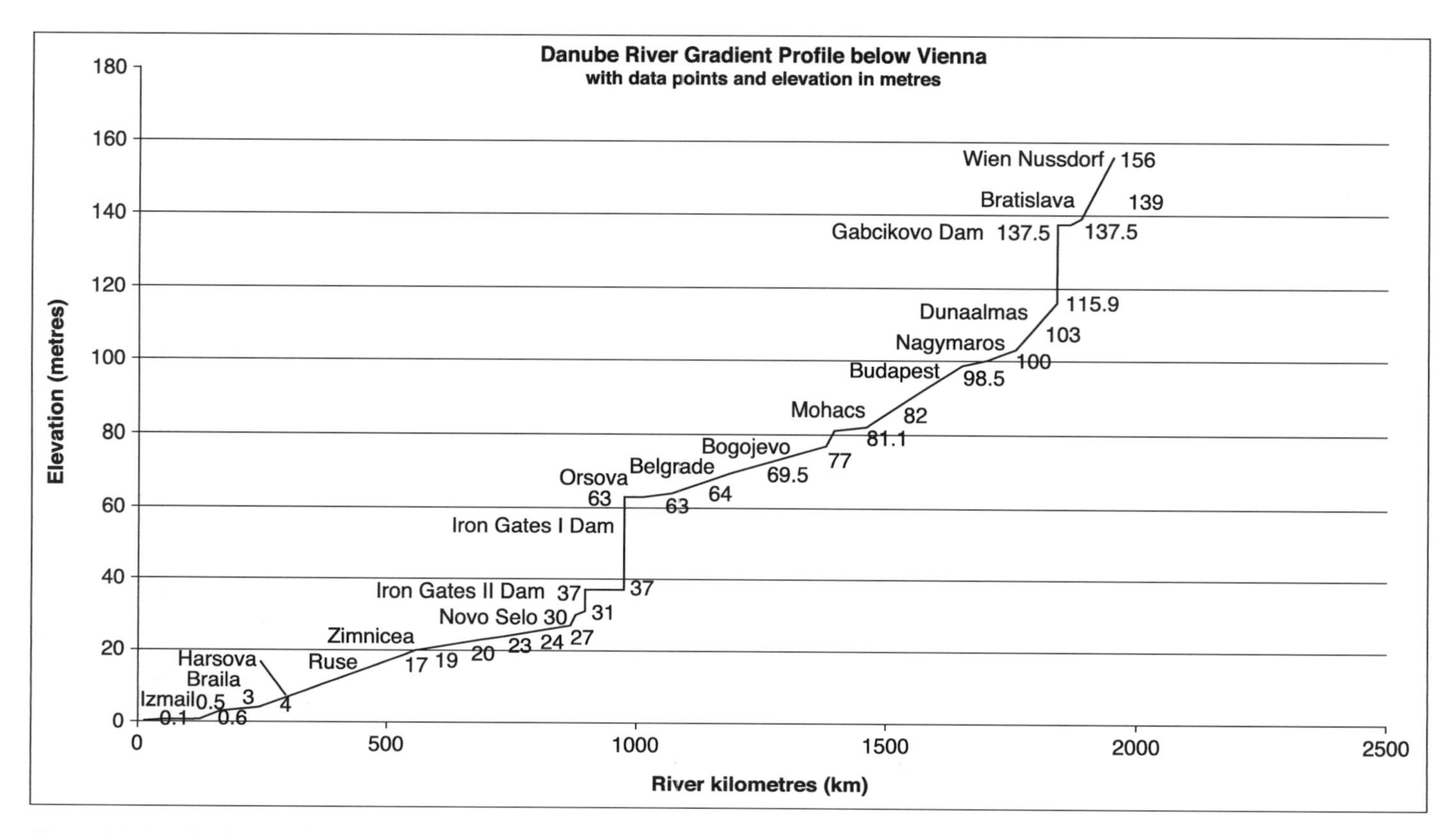

Figure 6.3 Danube longitudinal surface river gradient profile below Vienna.

FORMER DANUBE RIVER GRADIENT

The former Danube River gradient profile below Vienna (Figure 6.4), shows that this river gradient profile has been altered less by the two major HEP schemes constructed, the Iron Gates in 1970 and the Gabcikovo in 1992, than the other rivers in the sample. This is due to the 'run of the river' nature of these two schemes, with relatively small reservoir volumes and lengths, as well as the fact that the head heights in each case – 26 m and 7 m for Iron Gates I and II, and 21 m for the Gabcikovo scheme – are relatively modest compared to most of the sample.

COLUMBIA RIVER GRADIENT PROFILE FROM MCNARY RESERVOIR

The first 1,200 km of the Columbia River can be seen to be almost entirely converted into a cascade of dams and reservoirs (Figure 6.5). The average gradient from the upstream end of the McNary dam to the sea is 0.019 per cent as the river descends from 104 m to sea level over 540 km. However, the average gradient between the upper end of the McNary reservoir and just downstream of the Bonneville dam is steeper at 0.032 per cent, while the gradient of the last 233.7 km to sea level is just 0.0021 per cent. Some sections are still steeper, for example from the upper end of the Bonneville reservoir to the upper end of the Dalles reservoir, at 0.066 per cent. The reservoirs have inundated rapids sections at several points.

FORMER COLUMBIA RIVER GRADIENT PROFILE

It can be seen from the gradient profiles Figures 6.5 and 6.6 that the Columbia River has been changed very considerably by the construction of a series of 11 dams over the ~1,200 km course through the USA, to the extent that only two reaches, the Hanford reach and the last tidal reach to the sea, are in their natural state. The size of the large HEP scheme, the Grand Coulee, at the upper section, with its 105 m dam head height and 233 km long Rufus Wood Lake reservoir, with its 1.179×10^{10} m^3 capacity is striking.

COLORADO RIVER GRADIENT PROFILE

The Colorado River gradient profile below Lake Powell is shown in Figure 6.7. The average gradient in this section is 0.079 per cent, as the river falls 1,128 m over 1,431 km. However, the river gradient shows four or five different sections; the first 545 km from the Mexican border has an average gradient of 0.032 per cent; the next 423 km is steeper, with an average gradient of 0.127 per cent; following this the next section of 41 km gets steeper, to an average gradient of 0.22 per cent. The next 98.4 km has a gradient of 0.117 per cent, and after this the 323 km to the upper end of Lake Powell at kilometre 1,431 (from the Mexican border) is less steep, with an average gradient of 0.059 per cent.

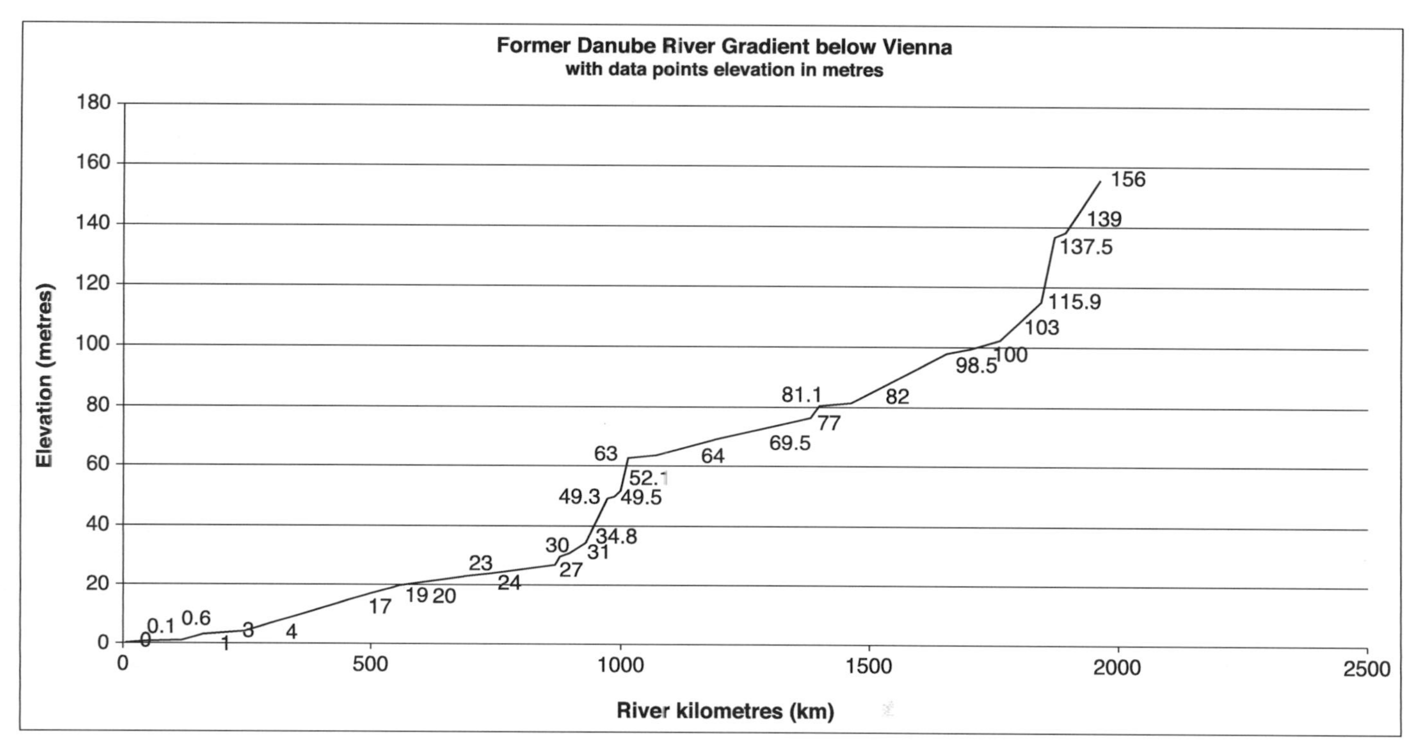

Figure 6.4 Former Danube River surface gradient profile below Vienna.

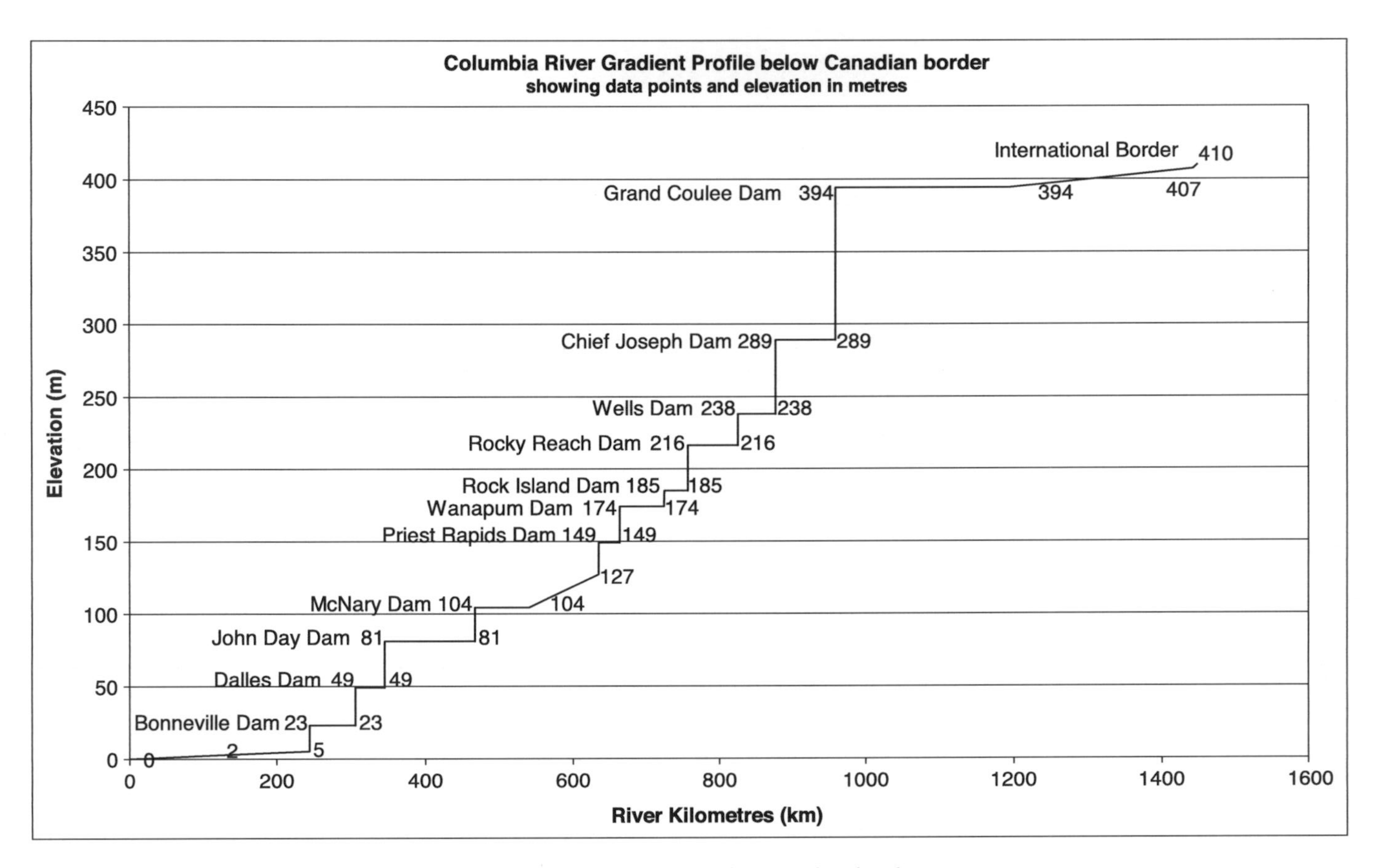

Figure 6.5 Columbia River longitudinal surface gradient profile below the Canadian border.

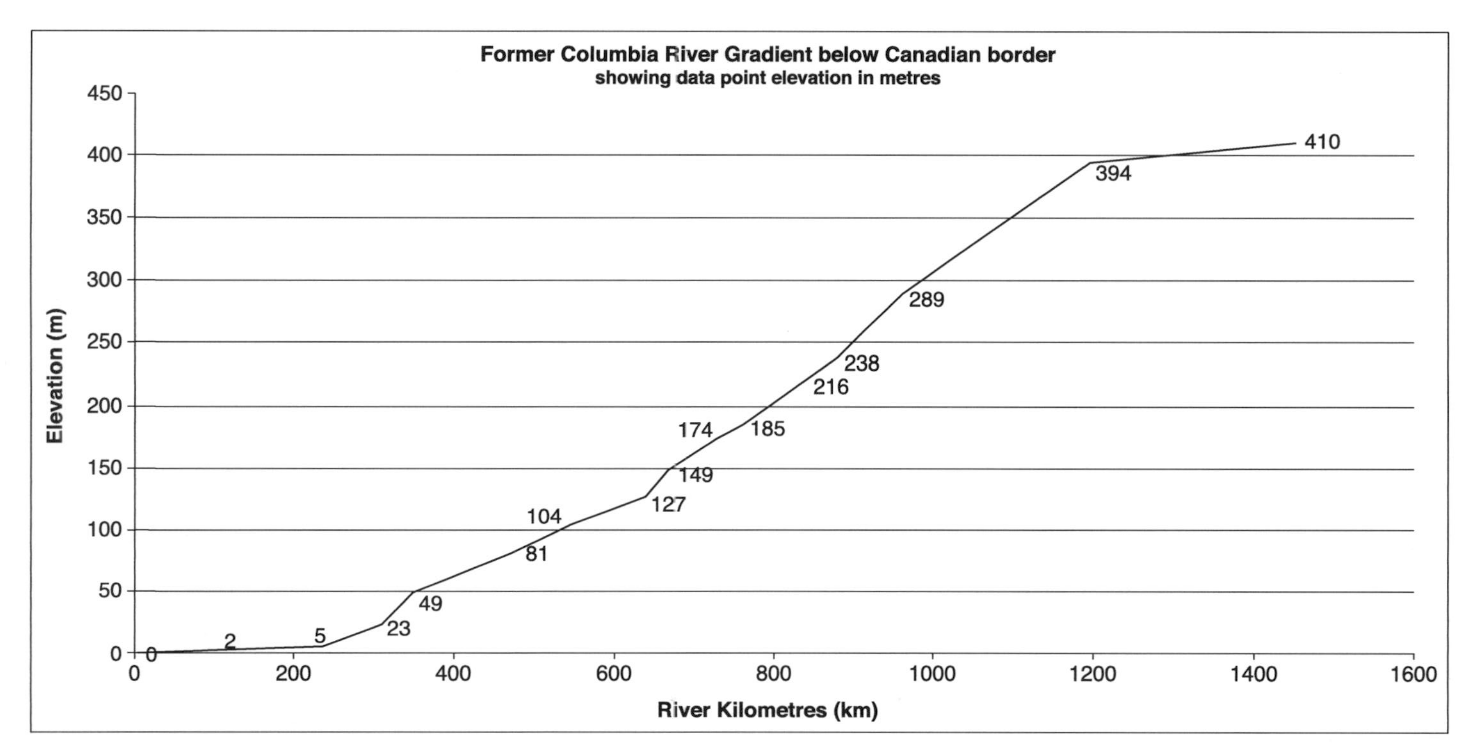

Figure 6.6 Former approximate Columbia River surface gradient profile below the Canadian border.

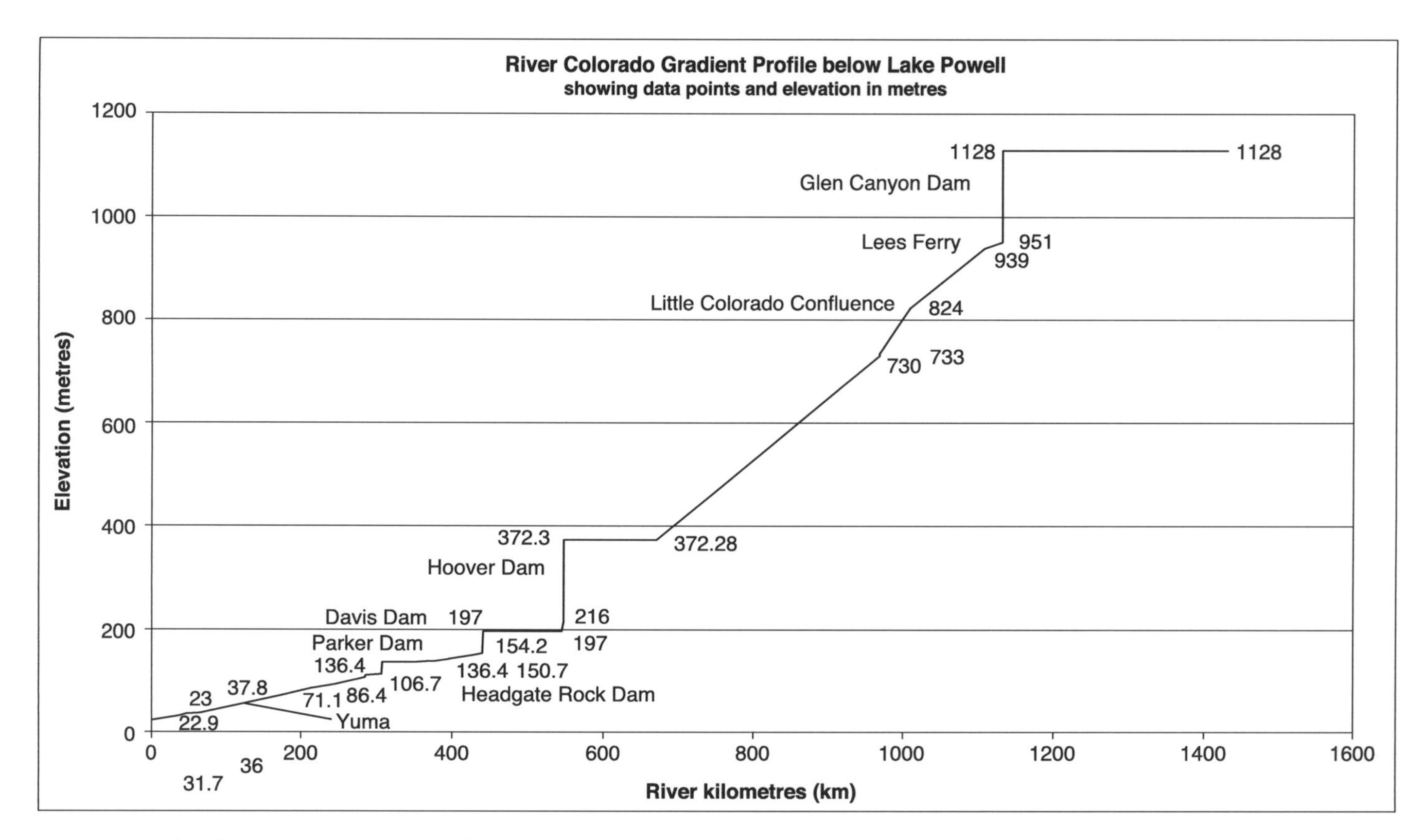

Figure 6.7 Colorado River approximate surface gradient profile below Lake Powell.

Although the Colorado River has two large HEP schemes in this lower 1,400 km section – the Glen Canyon scheme (built in 1964 with 155 m head height), and the Hoover dam (built 1933 and now with 158 m head height) – as well as three smaller dams – the Davis dam, the Parker dam and the Headgate Rock dam – the gradient profile is less changed than that of the Columbia. Between the Glen Canyon dam and the top of Lake Mohave (the reservoir for the Hoover scheme), there remains 461 km of natural river gradient, flowing through the Grand Canyon, with series of rapids.

Below the Headgate Rock dam at kilometre 286 (from the US border), the gradient profile is natural. The pre-existing gradient profile for the river in this section is shown in Figure 6.8.

6.2.2 The effect of the dams and reservoirs on flow rates downstream

The effect of the reservoirs downstream of large impoundment dams is generally to attenuate the variations between peak and minimum flows. Reduction of the peak flood flows will have great significance for the river's transport of sediment, since this is when most sediment transport occurs. The degree of attenuation will depend on the volume of the reservoir in relation to the river flow rate and on how the reservoir releases are managed, whether for electricity generation, water supply, navigation or flood control.

The attenuation of the variation between peak and minimum flows can be observed in Figure 6.9, showing the graph of the mean monthly flow rates on the River Nile at the old Aswan dam, just below the Aswan High dam, pre-1963 and post-1963. The flood flows from August to October, peaking in September, have been reduced from an average of ~8,700 $m^3 s^{-1}$ to less than 2,000 $m^3 s^{-1}$, as can be seen in Figure 6.10. The mean peak flow has now been moved to July. This is, of course, to be expected, since the purpose of the Aswan High dam was to store up to about two years of river flow and to be able to provide a constant water supply downstream for irrigation, domestic and industrial water and electricity generation. The Aswan High dam has been successful in these respects.

However, the loss of the flood flows, and the energy they represent, means that little sediment, ~2 per cent, is transported below the Aswan High dam (Marndouh 1985: 459).

6.2.3 The loss of power through attenuation of the flood flows

Since the instantaneous power of the river ultimately determines the extent of sediment 'pick up' and transport (Bagnold 1966), the difference in TSP in the peak flow period before construction of the dam and after can be estimated as a measure of the overall loss of stream power. The flow rates at Aswan, just below the Old Aswan dam and a little downstream from the High Aswan dam before the filling of the reservoir from 1967 to 1970 show how great the change is (see Figure 6.10).

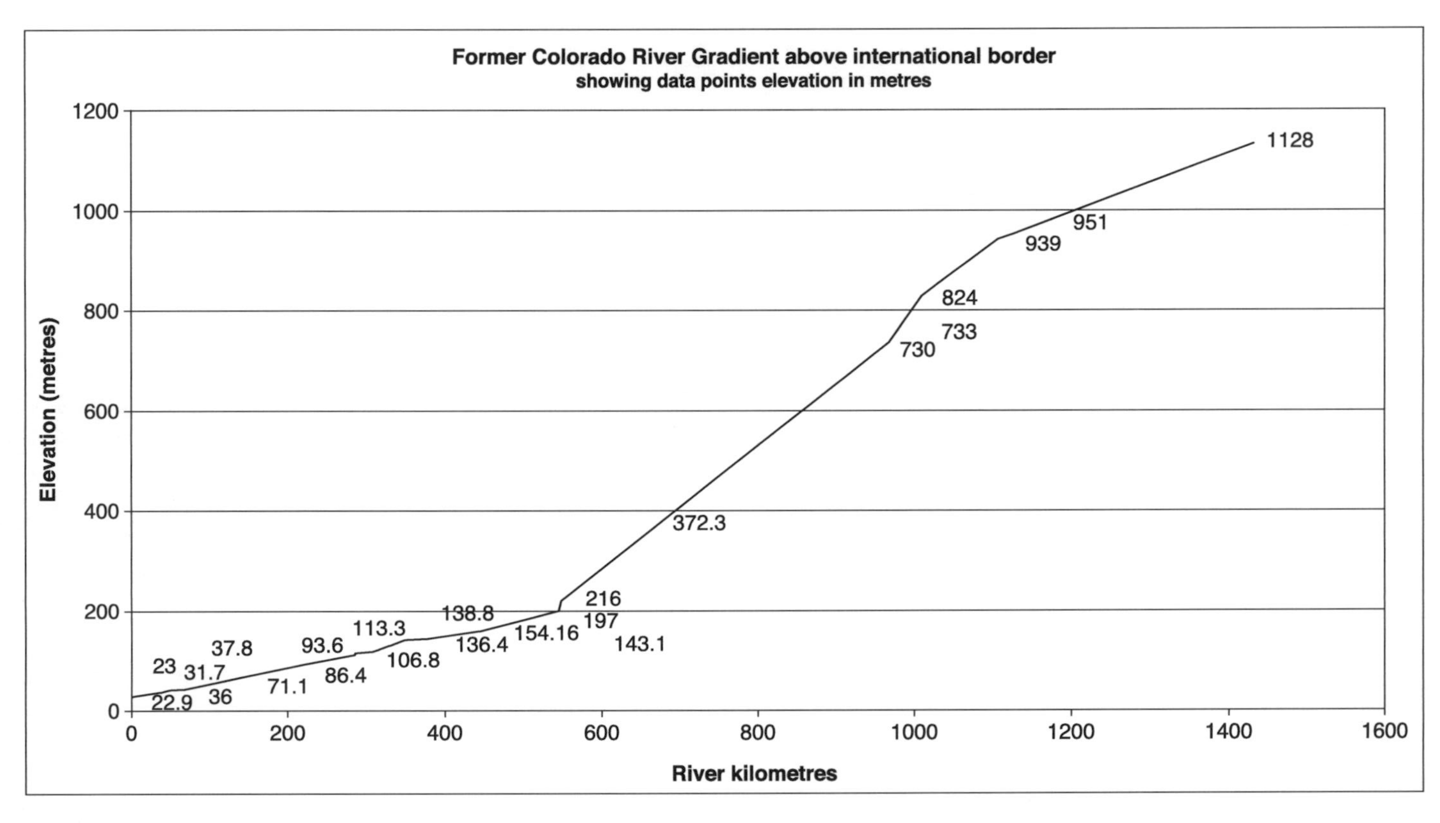

Figure 6.8 Approximate former Colorado River surface gradient profile.

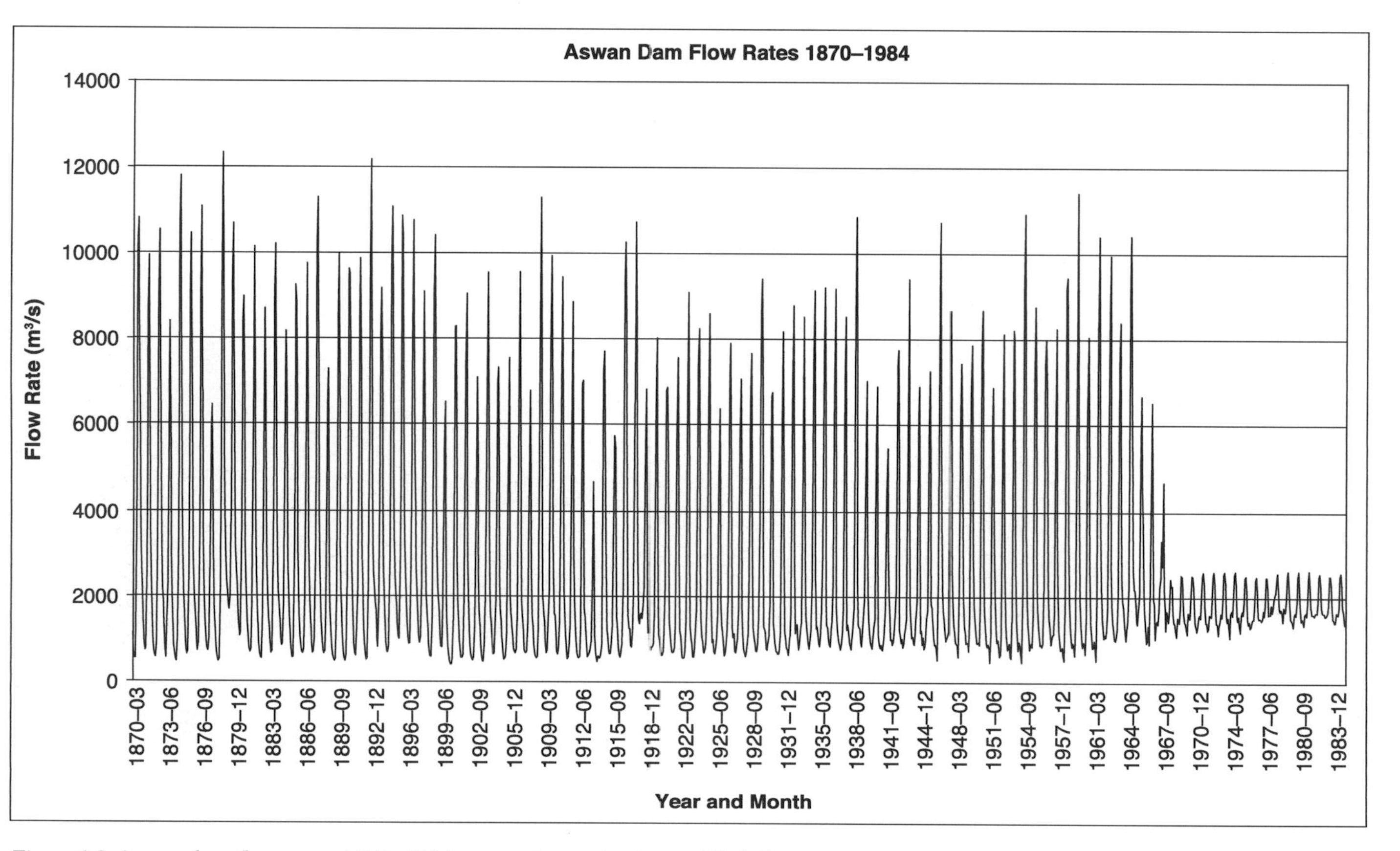

Figure 6.9 Aswan dam flow rates, 1870–1984, pre and post the Aswan High Dam.

Source: flow data from GRDC (2007/2008).

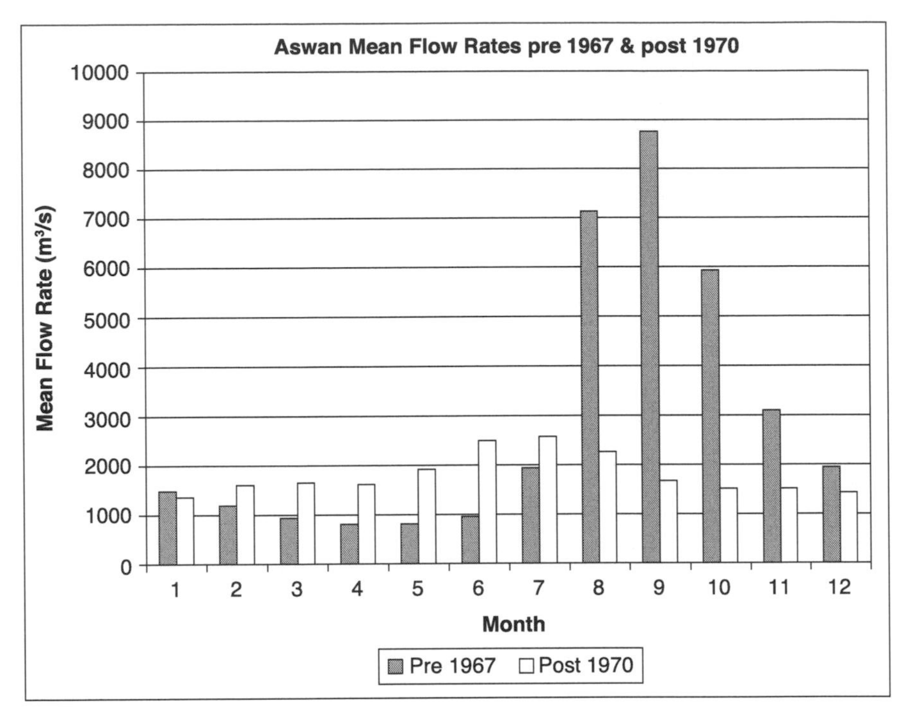

Figure 6.10 Aswan mean monthly flow rates pre-1967 and post-1970.

The peak mean monthly flow of 8,763 $m^3 s^{-1}$ in September, pre-filling of the reservoir from 1967, has been reduced to a peak mean monthly flow of 2,573 $m^3 s^{-1}$ in July, a reduction of the peak flow by over 70 per cent. The former minimum mean monthly flow of 812 $m^3 s^{-1}$ in April, pre-1967, has increased to a new minimum mean monthly flow of 1,365 $m^3 s^{-1}$ in January. A maximum September mean flow of 12,345 $m^3 s^{-1}$ for the period 1871 to 1967 has been reduced to 2,554 $m^3 s^{-1}$ in June for the period 1970–84. The hydrograph for the flow station has changed markedly as a result of the flow regulation. This change can be represented in TSP for the reach, i.e. in kilowatts.

It can be seen from Figure 6.11 that the TSP for the 35.7 km river reach Aswan to Gaafra has been reduced for the peak flow months of August to October and increased for the minimum flow months, in concert with the mean flow rate changes. From a peak value of 730,000 kW in September in the period 1871–1967, the TSP peak is 214,000 kW in July for the period 1970–84, since the dam construction. This is just 29 per cent of the previous peak TSP, a drop of 71 per cent.

This diminution of TSP is that for the reach *below* Aswan of 35.7 km, with a mean gradient of 0.0237 per cent, and is distinct from the TSP lost to the Aswan High dam and Old Aswan dam over the reservoir reach.

Data for the pre-dam period for points further down the river were not available, only post-1973, so it was not possible to estimate the TSP reduction

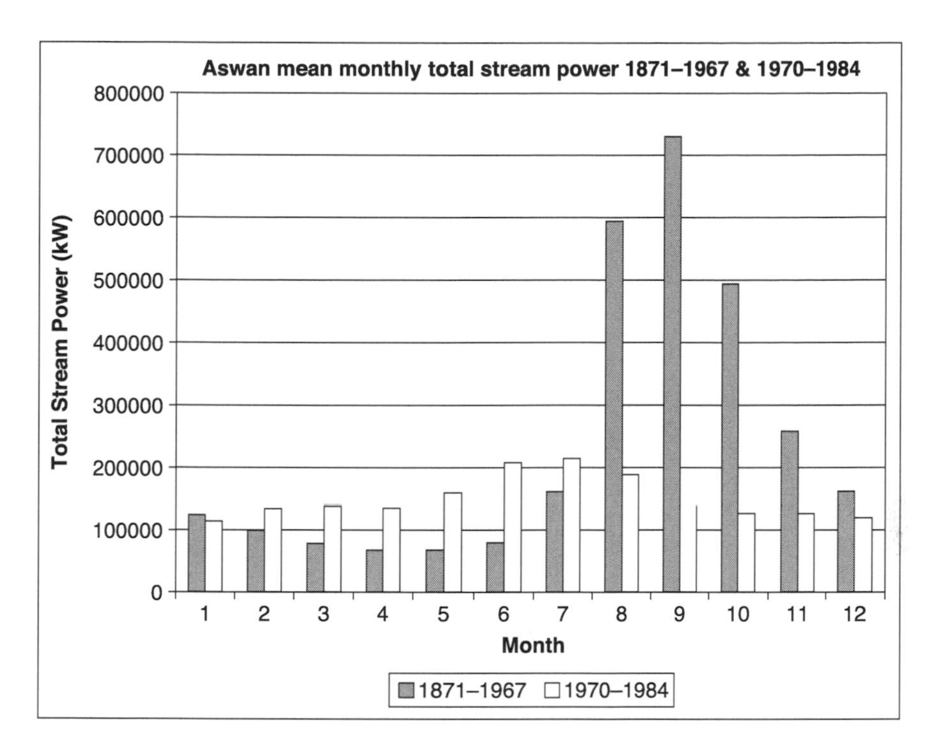

Figure 6.11 Aswan mean monthly total stream power, 1871–1967 and 1970–84.

for all the reaches below Aswan. However, the flow regime in Figure 6.11 could be expected to be reflected at points downstream, to the delta and sea mouth. This loss of TSP available for the picking up and transport of sediment is evident.

Danube mean flow rates post large dam construction

A certain amount of attenuation of the peak and minimum flows can be discerned on the two Danube HEP schemes studied here post-1970 below the Iron Gates scheme and below the Gabcikovo scheme post-1991. Figures 6.12 and 6.13 show the flow rates at Zimnicea measuring station, some 336 km below the Lower Iron Gates dam, and at Dunaalmas station 83 km below the Gabcikovo HEP scheme.

However, as can be seen, the effect on flow rates is slight, reflected in a diminution of maximum rates of flow from 14,000 $m^3 s^{-1}$ to ~12,000 $m^3 s^{-1}$, and with the frequency of such maxima reduced slightly. This effect can be seen in the reduction in standard deviation of flow rates pre-1970 and post-1970, in the peak months of April and May, the peak flow months, from 2,554 to 1,988 $m^3 s^{-1}$ (in April), and from 2,594 to 1,486 $m^3 s^{-1}$ (in May). The significance of this can be estimated in terms of reduction in stream power. The reduction in stream power for the two peak months of April

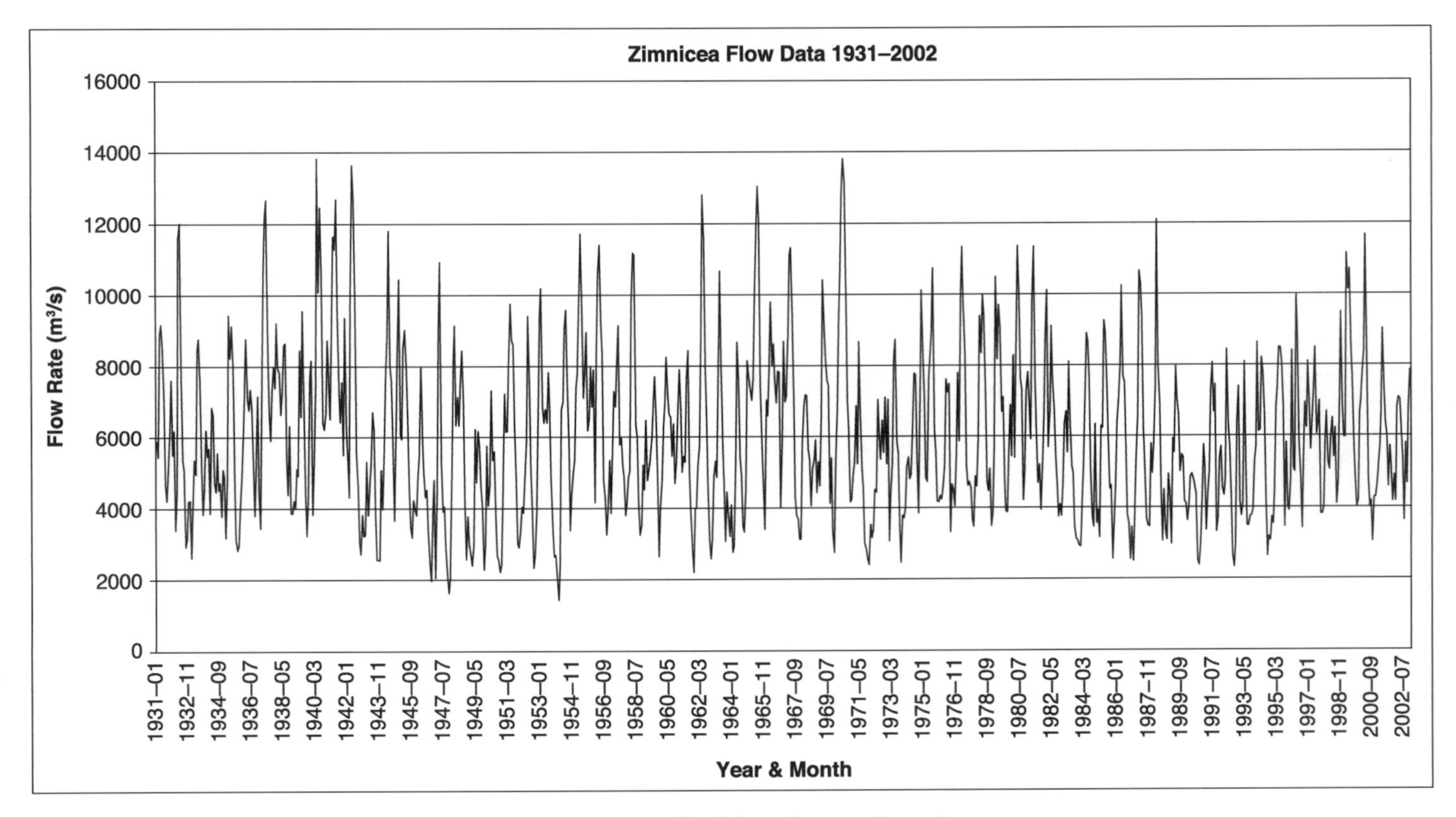

Figure 6.12 Zimnicea flow rates, Danube River 1931–2002, 336 km below the Iron Gates dam.

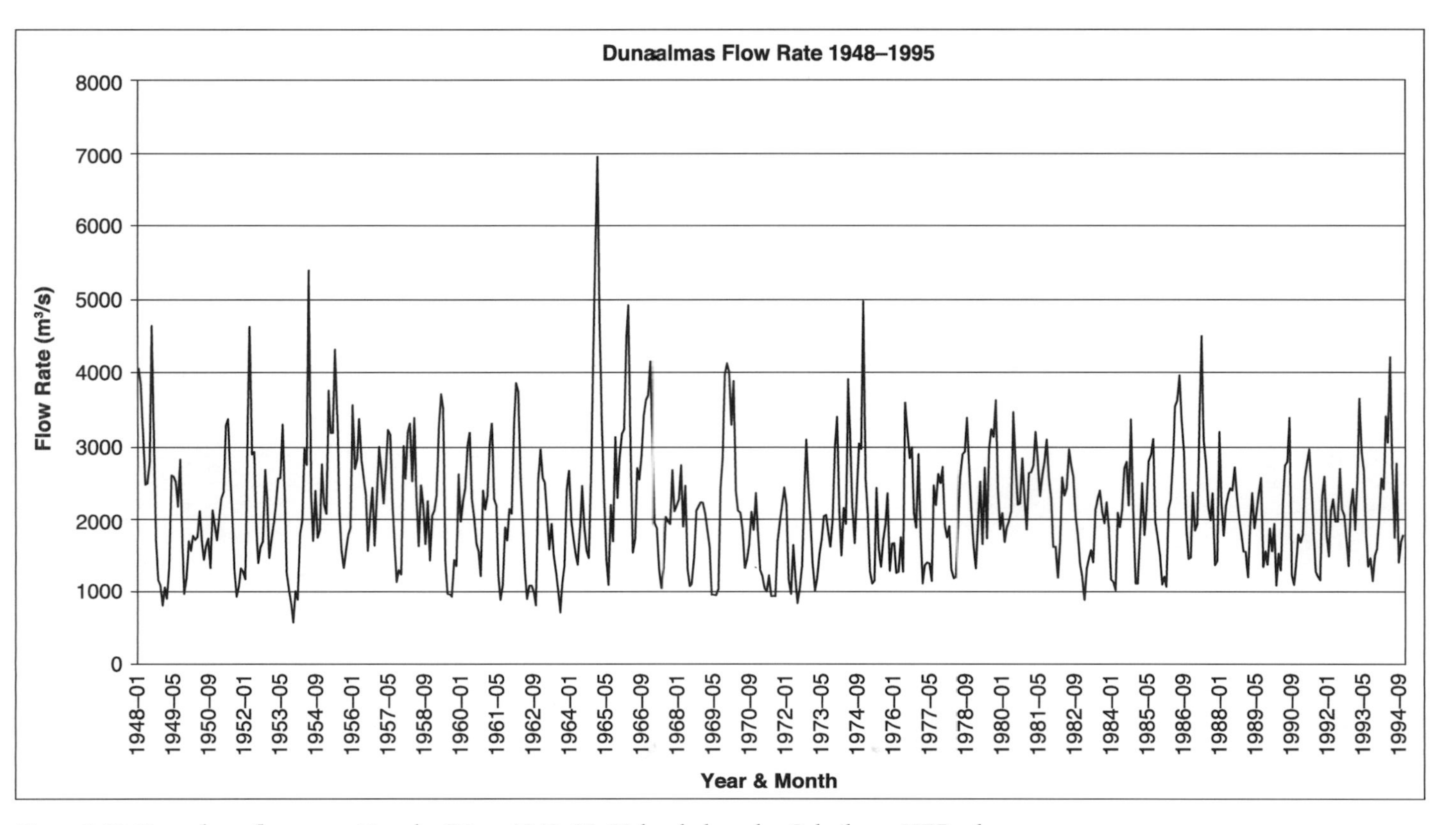

Figure 6.13 Dunaalmas flow rates, Danube River 1948–95, 83 km below the Gabcikovo HEP scheme.

and May, after the dam construction, has been estimated for the reaches between the flow measuring stations of Novo Selo, 21 km below the lower Iron Gates dam, and for Lom (157.6 km below the lower Iron Gates dam) and then for the Lom–Jiul reach, on the Danube. For the 128.5 km Novo Selo to Lom reach, the reduction in TSP for the peak months of April was from 244,193 kW to 227,442 kW, or 6.9 per cent, and for May from 230,000 kW to 206,000 kW, or 10.4 per cent. TSP for the 51 km reach from Lom to Jiul has been calculated as 81,100 kW for April before dam construction, and 76,100 kW after dam construction, a reduction of 6.14 per cent. For May a reduction from 76,000 kW to 69,800 kW equates to a reduction of 8.14 per cent. The reduction in stream power is not great and unlikely to be significant given the variation in flow.

At Dunaalmas, 83 km below the dam, attenuation of flood flows is also relatively slight, and the effect of the Gabcikovo dam in 1992 is not apparent (Figure 6.13). However, the data set here runs only until 1995, some three years after the Gabcikovo scheme was operating, and therefore cannot be a representative sample. The gradual diminution of the peak flows between 1948 and 1995 may be the result of the cumulative damming of the river upstream, where numbers of dams have been constructed since the 1960s (Schwarz 2008).

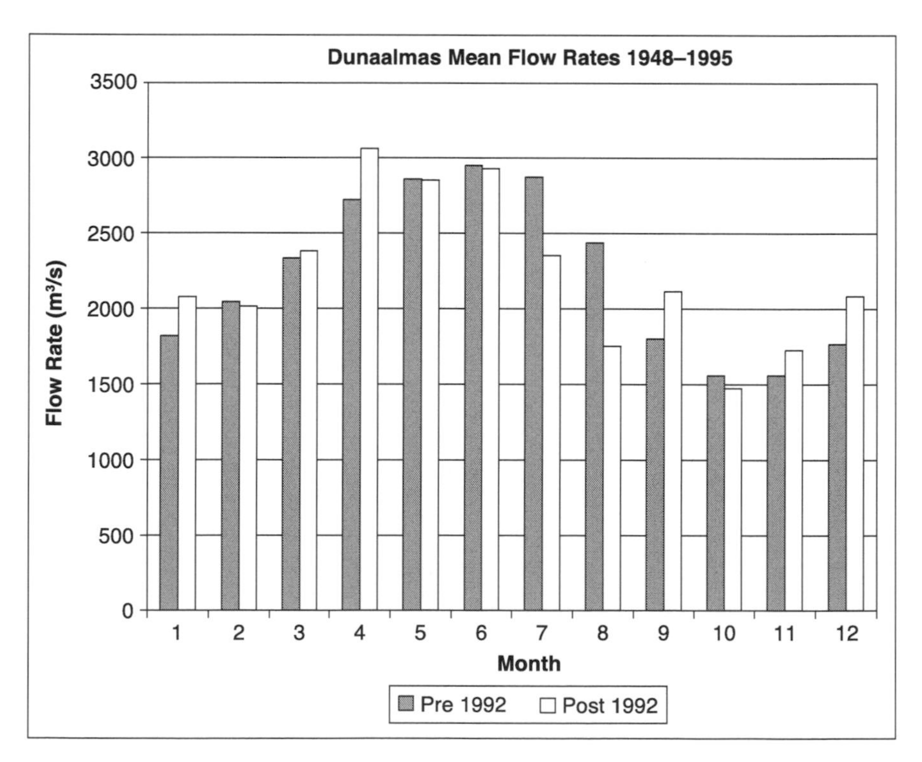

Figure 6.14 Dunaalmas mean monthly flow rates 1948–95, pre-1992 and post-1992.

Little or no discernible effect can be seen in the mean monthly hydrograph flow rates for Dunaalmas, 87 km below the Gabcikovo HEP scheme (Figure 6.14), allowing for the small amount of data available, just three years following dam completion.

Columbia River flow rate changes since dam construction

The Columbia River has, as stated before, been extensively impounded by large dams in the upper and middle reaches, as have its tributaries, for example the Snake River. In particular the large Canadian Mica dam of 1976 and the Grand Coulee dam of 1941 have attenuated the flood and low flows.

The effect on the flow rates by the storage dams upstream of the Grand Coulee dam on the Columbia River can be seen in Figure 6.15, with the diminution of the variation in flow rates. This is especially marked post-1973, after the building of the Mica dam in Canada in that year.

The effect of the three Canadian storage dams on the Columbia River flow can be seen in Figure 6.16, which shows the flow rates pre- and post-1973. Prior to the Canadian storage dams, the flood flows were very pronounced from May to July, exceeding 5,000 m^3 s^{-1}, with a peak of over 8,000 m^3 s^{-1} in June itself, compared to the mean flow of 3,100 m^3 s^{-1} (GRDC 2008). Subsequently there is only a minor peak in June and a second slight peak in January, formerly the lowest-flow month.

This attenuation of the summer flood flow continues downstream throughout the river, and can be calculated in TSP terms to provide before- and after-dam TSP value changes.

In Figure 6.17, the diminution of flood flows after 1973 can be seen at the flow measuring station below Rock Island dam, 209 km below Grand Coulee dam.

The peak flow rate has diminished by approximately half, and the minimum flows in October are slightly increased, as shown in Figure 6.18. Further down the river, 323 km below the Grand Coulee dam, the effects of the Mica dam impoundment can be observed. See Figure 6.19 for the TSP at the measuring station below Priest Rapids for the years 1917–73, and then 1973–2006. The flood flow TSP has diminished by half in June, from 2,050,000 kW to 1,014,000 kW, while the low flows have increased by one and a half times from 368,000 kW in January to 514,000 kW in November for the 95 km reach from below Priest Rapids to the upstream end of the McNary reservoir. The diminution in variation can be expressed as a reduction in maximum mean monthly flows from 16,539 m^3 s^{-1} in June between 1917 and 1973, to 9,031 m^3 s^{-1} between 1973 and 2006, and an increase in the minimum mean monthly flows of 607 m^3 s^{-1} in January to 1,621 m^3 s^{-1} in April.

At the Dalles flow station at 847 km below the Grand Coulee, complete flow rate data from 1878 to 2006 (Figure 6.20), shows the diminution of variation both since 1940 and, more pronounced, since 1973. The Dalles is situated at 306 km from the river mouth, 656 km below the Grand Coulee

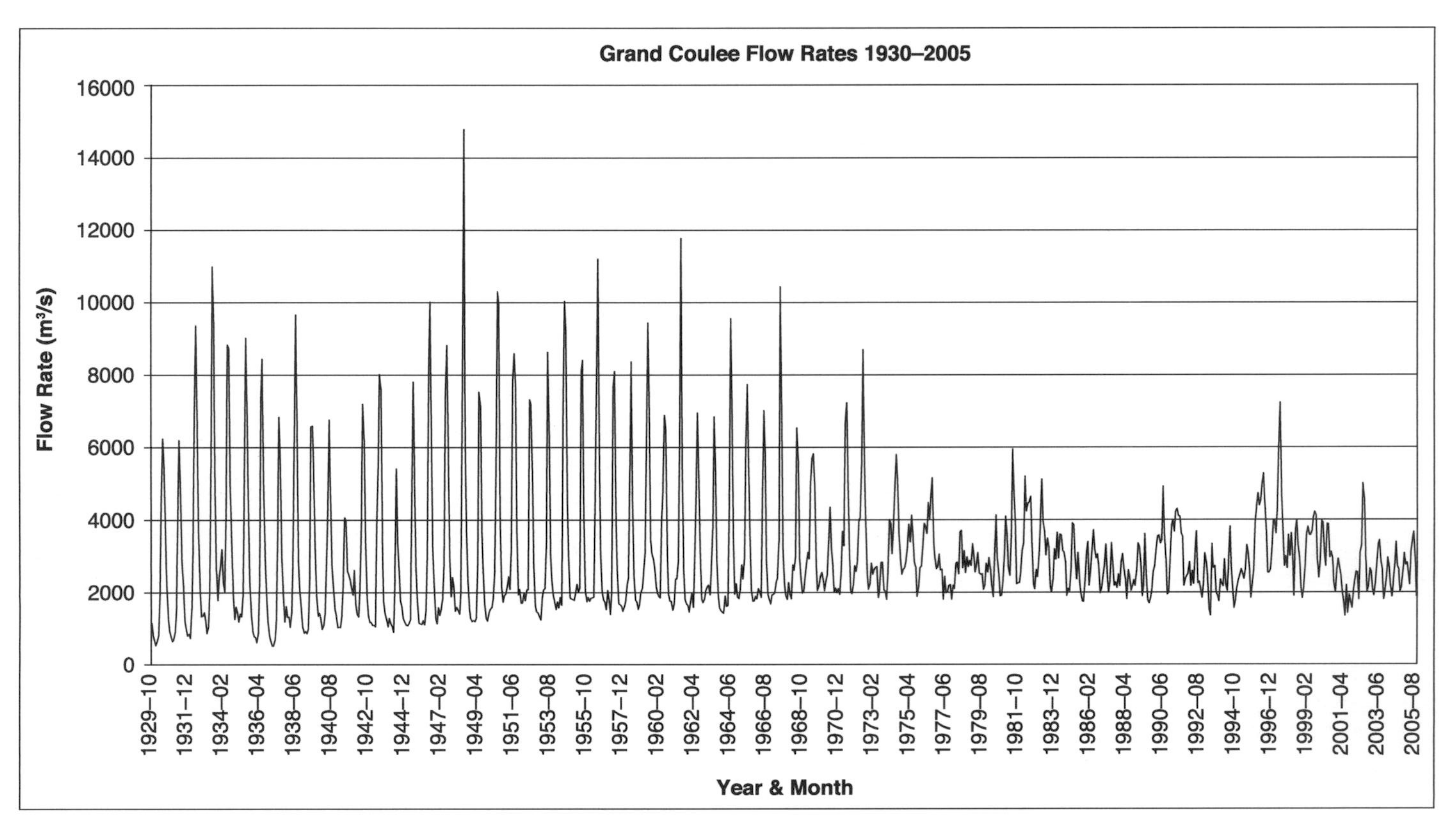

Figure 6.15 Grand Coulee flow rates 1930–2005.

Source: adapted from GRDC data (2008).

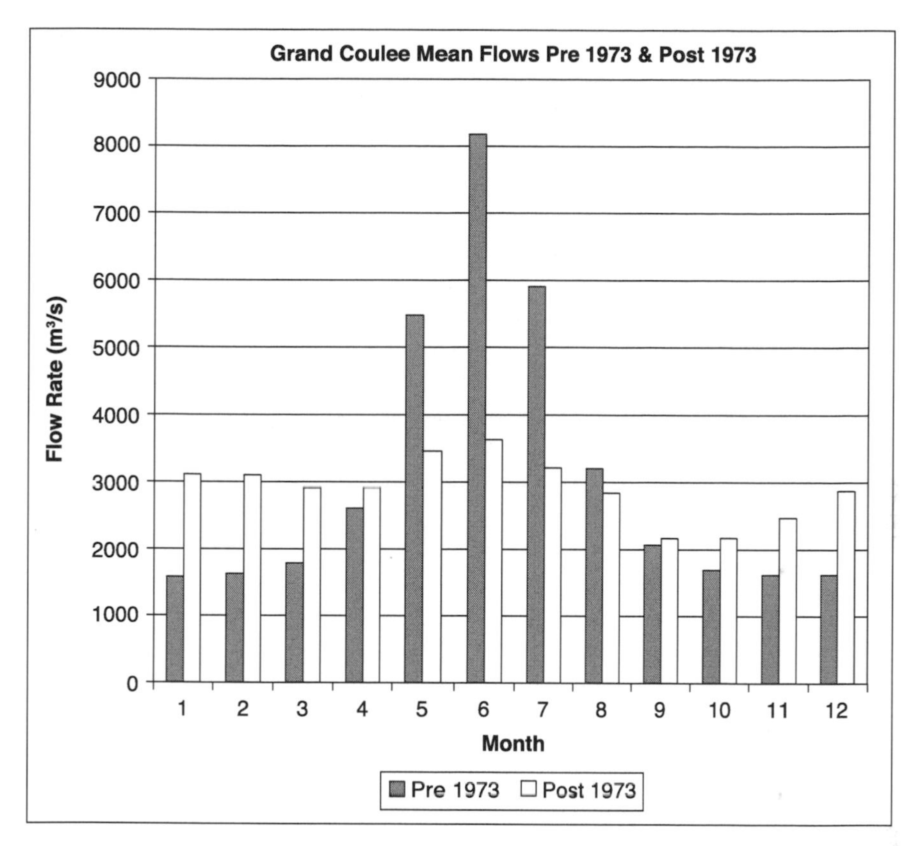

Figure 6.16 Monthly mean flow rates Grand Coulee, pre-1973 and post-1973.

Source: adapted from GRDC data (2008)

dam, and it is approximately 194 km below the confluence with the largest tributary, the Snake River, which has also been subject to impoundment.

Once again a diminution of the peak flow by approximately half from 13,879 $m^3 s^{-1}$ to 7,520 $m^3 s^{-1}$ is seen from Figure 6.21, together with a shift in the peak from June to May, and with low flows increased by about from 2,814 $m^3 s^{-1}$ in January to 3,174 $m^3 s^{-1}$ in September. The maximum flow has been reduced from 28,377 $m^3 s^{-1}$ in the period 1878–1972, to 13,636 $m^3 s^{-1}$ for the period 1973–2006. Minima are increased from 1,187 $m^3 s^{-1}$ to 2,121 $m^3 s^{-1}$ in these periods. While the maximum flows still occurred in the same month, i.e. the former peak flow month of June, the minima are transferred from January to September.

The standard deviation for the peak flow period had decreased from 3,941 $m^3 s^{-1}$ to 2,581 $m^3 s^{-1}$ in the 1973–2006 period, i.e. the variability had decreased. For the former minimum flow period in January, the standard deviation increased from 961 $m^3 s^{-1}$ to 1,060 $m^3 s^{-1}$, i.e. the variability had increased.

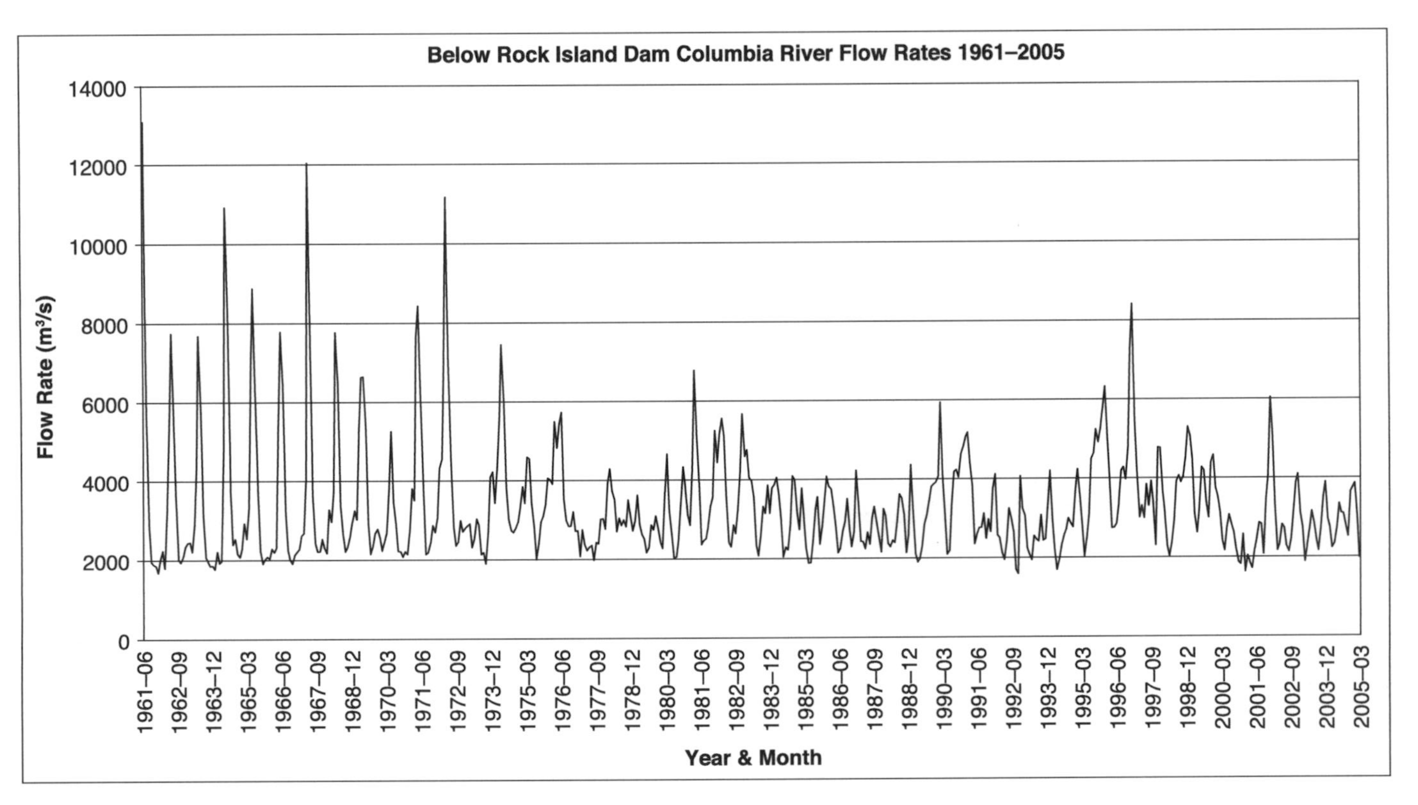

Figure 6.17 Flow rates for below Rock Island dam, Columbia River 1961–2005.

Source: adapted from GRDC data (2008)

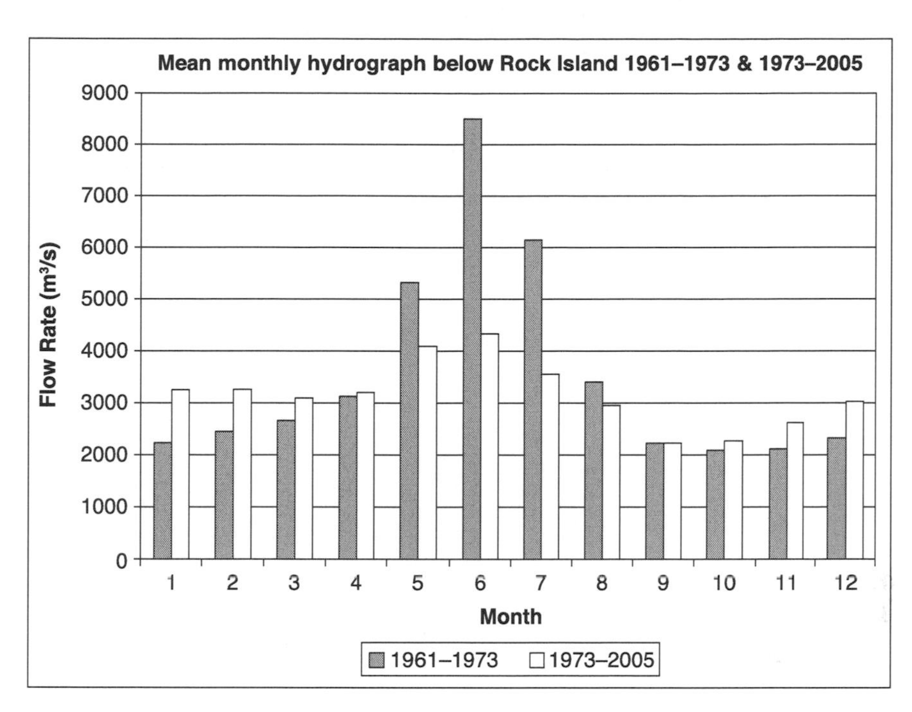

Figure 6.18 Mean monthly flow rate hydrograph below Rock Island 1961–73 and 1973–2005.

Source: adapted from GRDC data (2008).

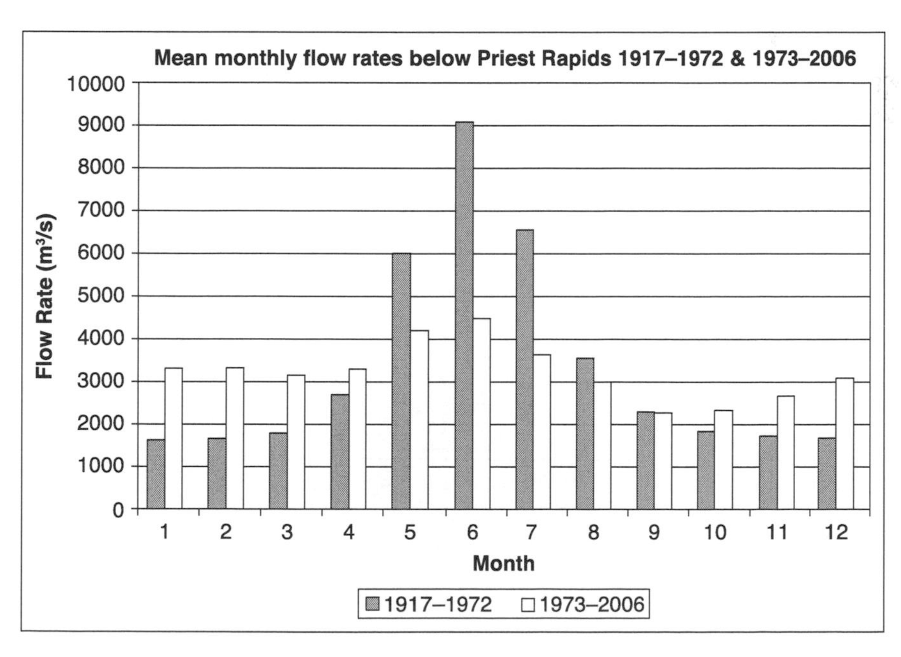

Figure 6.19 Mean monthly total stream power below Priest Rapids 1917–72 and 1973–2006.

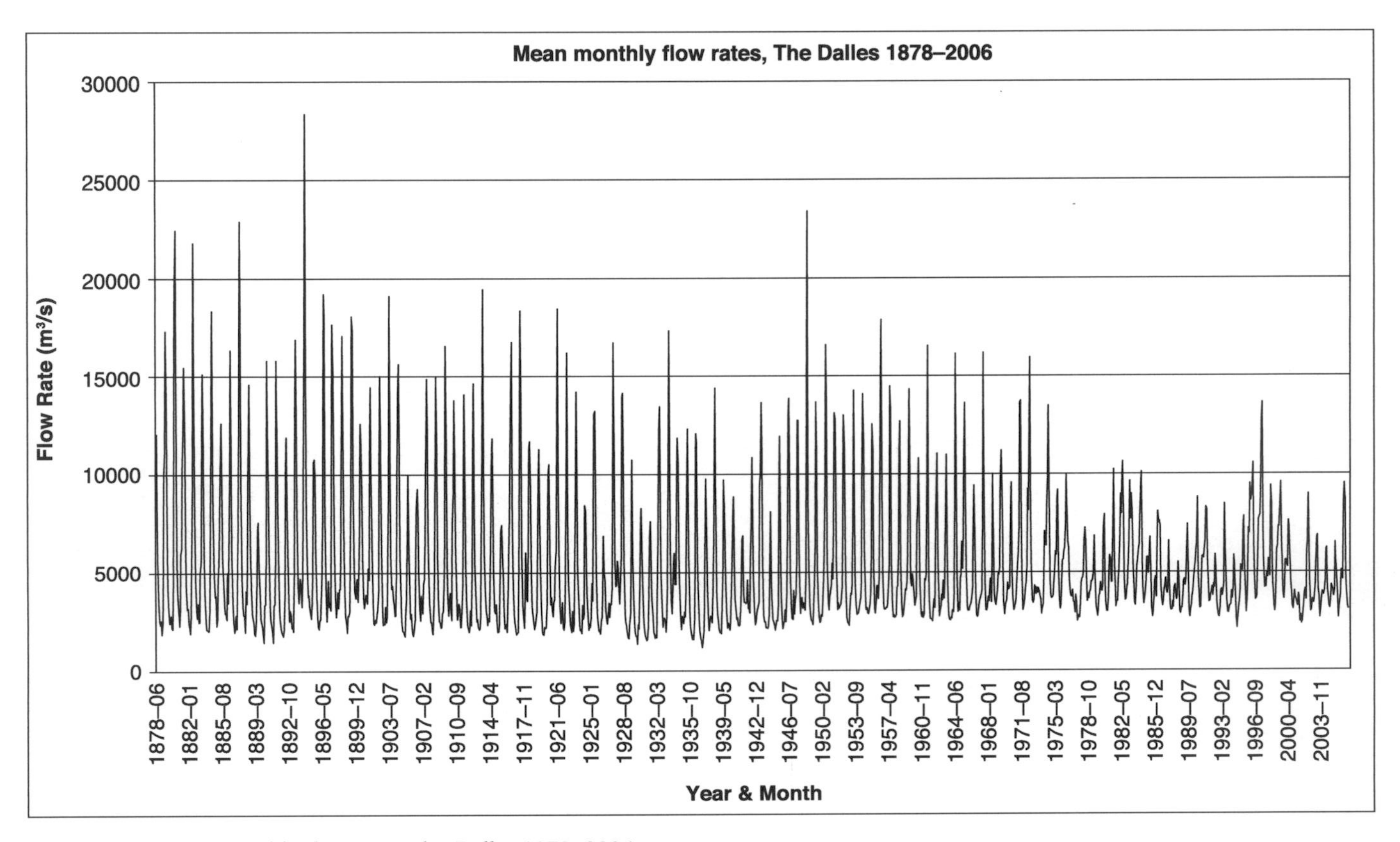

Figure 6.20 Mean monthly flow rates, the Dalles 1878–2006.

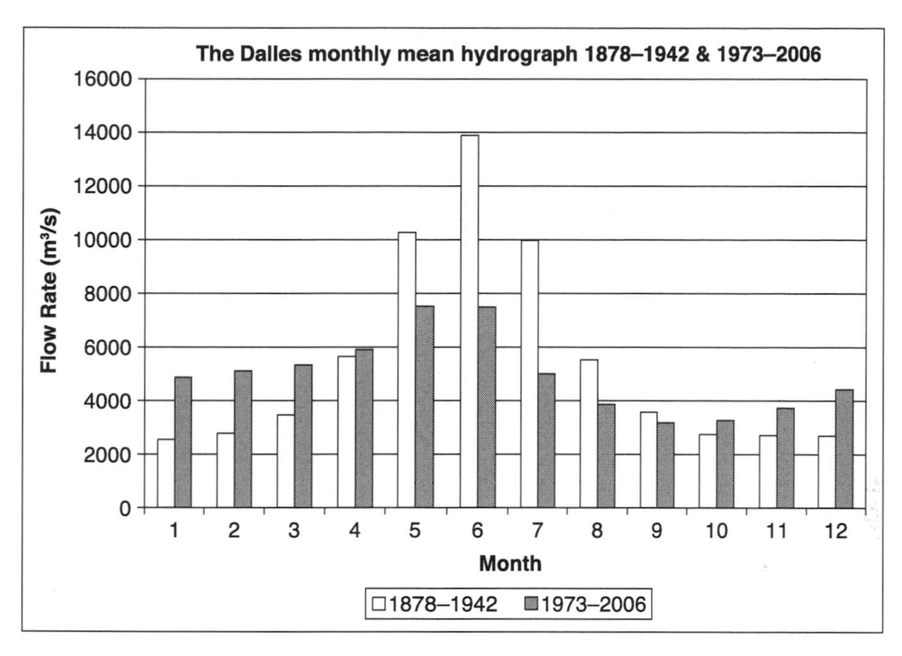

Figure 6.21 Mean monthly hydrograph for the Dalles, Columbia River 1878–1942 and 1973–2006.

Source: adapted from GRDC data (2008).

Colorado River flow rate changes and effects of impoundment and water abstraction on TSP

Although the Colorado River has been less impounded than the Columbia River, it flows with great seasonal flow variations, through an arid region with few major tributaries, and where much of the water is abstracted for irrigation and water supply. The effect of the large dams on the flow rates has been very pronounced as they have been designed to store about one year's river flow in the case of the Glen Canyon scheme and 2.8 years' flow in the Hoover dam scheme.

The changes in flow rates from 1911 to 2003 can be seen in Figure 6.22 for Lees Ferry, 24 km below Glen Canyon dam on the Colorado River after the construction and filling of the dam in 1963–4. Very low flows of mostly between 28 and 32 m^3 s^{-1} can be observed from early 1963 to early 1964 while the reservoir was filling.

As is clear from the flow rate graph and the mean monthly hydrograph in Figure 6.23, the flow regime has been considerably altered, with a near elimination of the flood and low flows, apart from some high flows in 1983 to 1986, peaking each year at over 1,000 m^3 s^{-1}, when presumably the reservoir capacity was not sufficient to contain these high flows.

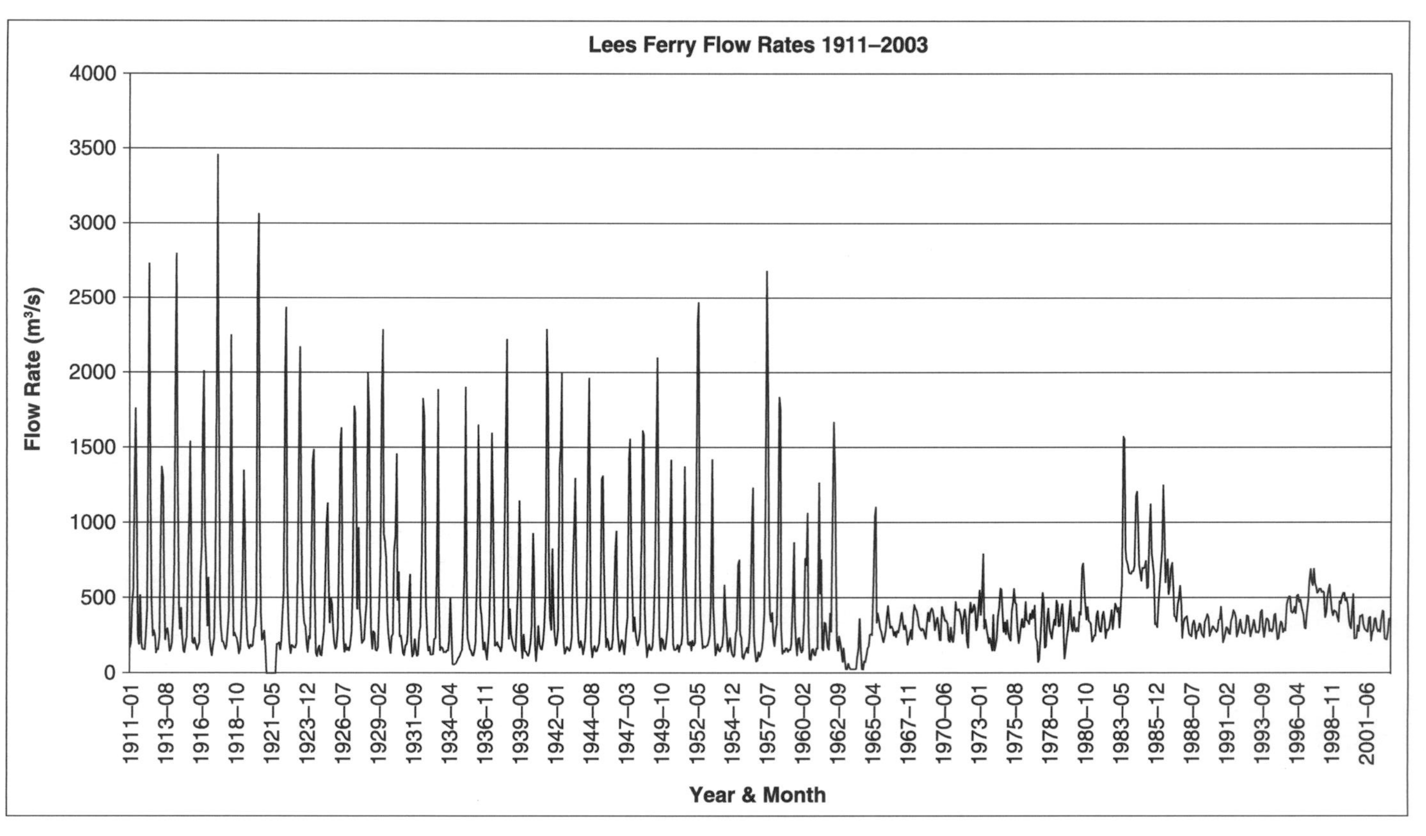

Figure 6.22 Lees Ferry flow rates 1911–2003, below Glen Canyon dam, Colorado River.

Source: adapted from GRDC data (2008).

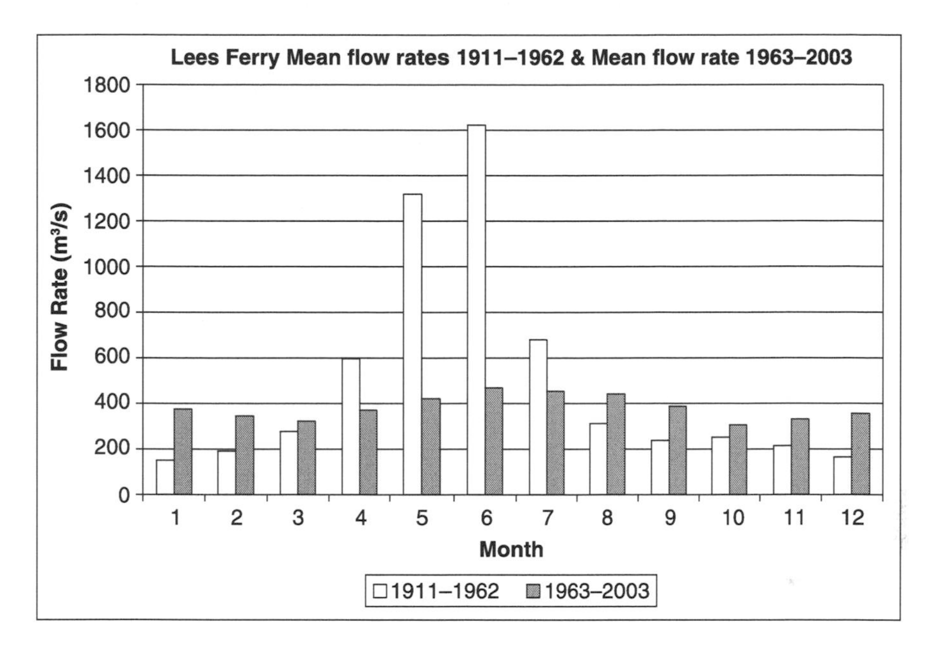

Figure 6.23 Mean monthly flow rates at Lees Ferry, Colorado River below Glen Canyon dam, pre- and post-1963.

The loss of seasonal flow variation is pronounced below the Glen Canyon dam on the Colorado River, with maximum flows reduced by almost a factor of four, from 1,600 m^3 s^{-1} to a little over 400 m^3 s^{-1} (see Figure 6.23).

There is a reduction of TSP in proportion to the loss of peak flows and low flows after the construction of Glen Canyon dam (Figure 6.24).

Even more outstanding are the changes in mean flow rate records at Yuma, 1,038 km downstream from Glen Canyon dam on the Colorado River in Arizona, near the Mexican border, and 500 km downstream of the Hoover major impoundment dam, for the period 1904–83 (Figure 6.25). The river flow rates show the effect of the Hoover dam and reservoir in 1933, with associated water abstraction.

After 1959, increased water abstraction led to a drop in mean flows from 612.96 m^3 s^{-1} in the period 1904–33, to 187.3 m^3 s^{-1} between 1934 and 1959, and then just 23.4 m^3 s^{-1} in the period 1960–75. By the late 1980s flow rates were averaging about 18 m^3 s^{-1} (IBWC 2001), though since the 1990s slightly increased flow rates have been enabled. Not only are the flood peak flows almost entirely evened out, but some 96 per cent of the former flow has been abstracted. With the loss of water flow, there is of course a loss of TSP.

The loss of TSP at Yuma Arizona on the Colorado River can be seen in Figure 6.26. From a mean TSP of 25,851 kW, the total available stream power has diminished to just 987 kW. There has been a loss of the order of 96.18 per cent in the mean TSP. Even more important than the mean TSP values is the diminution from a peak mean monthly TSP of 82,003.75 kW

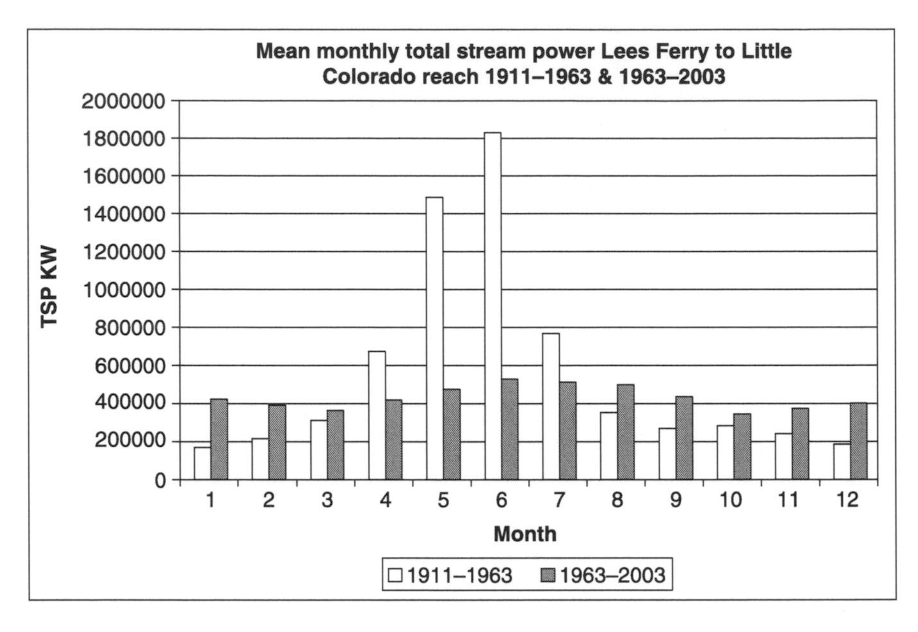

Figure 6.24 Mean monthly total stream power, for the Lees Ferry to Little Colorado reach, 1911–63 and 1963–2003.

in June before 1934, to a peak of just 1,181.12 kW in May, after 1960. This represents just 1.4 per cent of the former flood flow TSP, and so 98.6 per cent of TSP has been lost. The effects of this abstraction of water and hence TSP on the Colorado River are evident, in that very little sediment is transported to recharge the former delta at the mouth. This feature is now in greater part a residual one, and is just one of the environmental impacts resulting, though perhaps the most significant.

6.3 Land use

The land use in TWh km^{-2} p.a. for the sample HEP schemes was compared to the parameters:

- hypothetical power flux density derived from head height
- reservoir reach TSP
- CSP
- average water residence time
- former mean river gradient
- reservoir volume
- reservoir average fluid velocity
- production factor
- mean river flow.

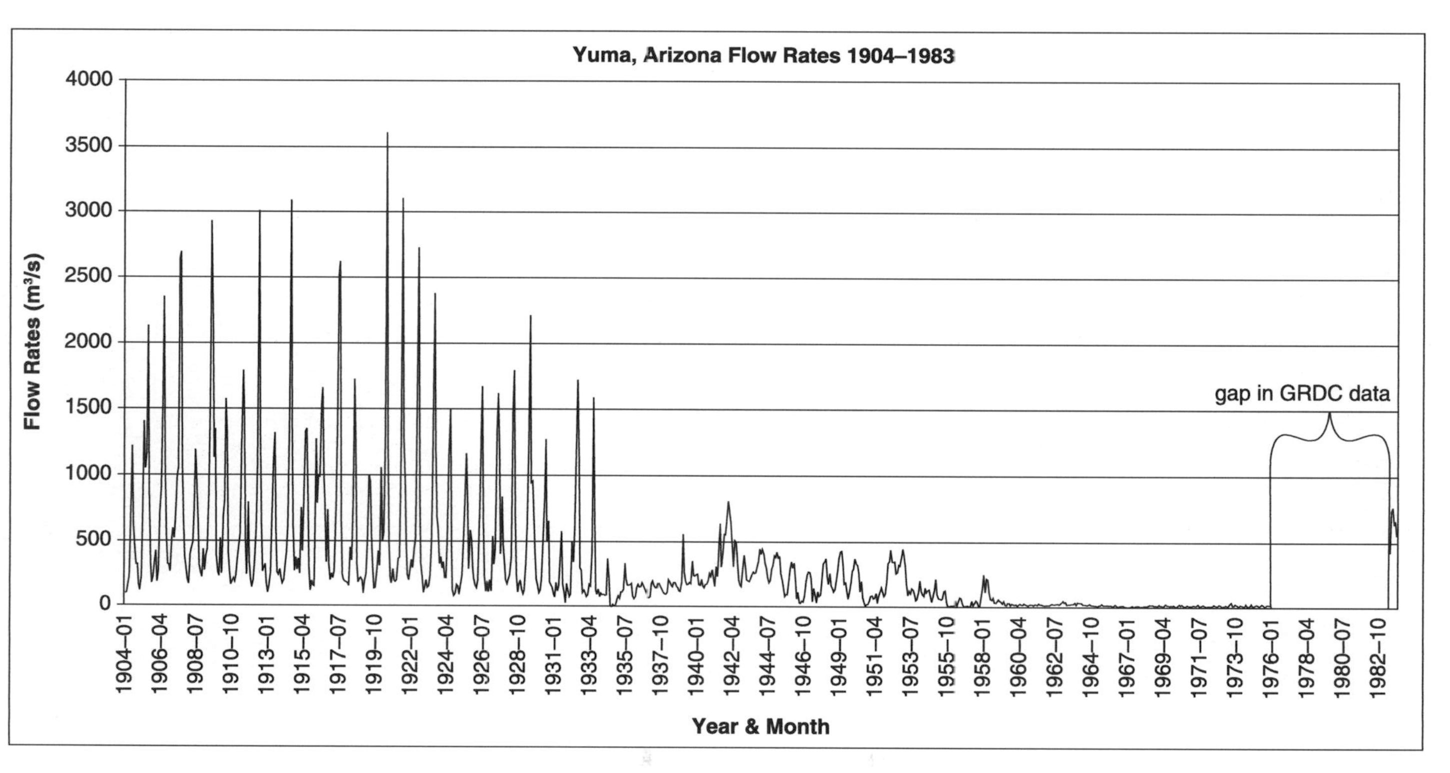

Figure 6.25 Flow rates at Yuma, Arizona 1904–1983, Colorado River.

Source: adapted from GRDC data (2008).

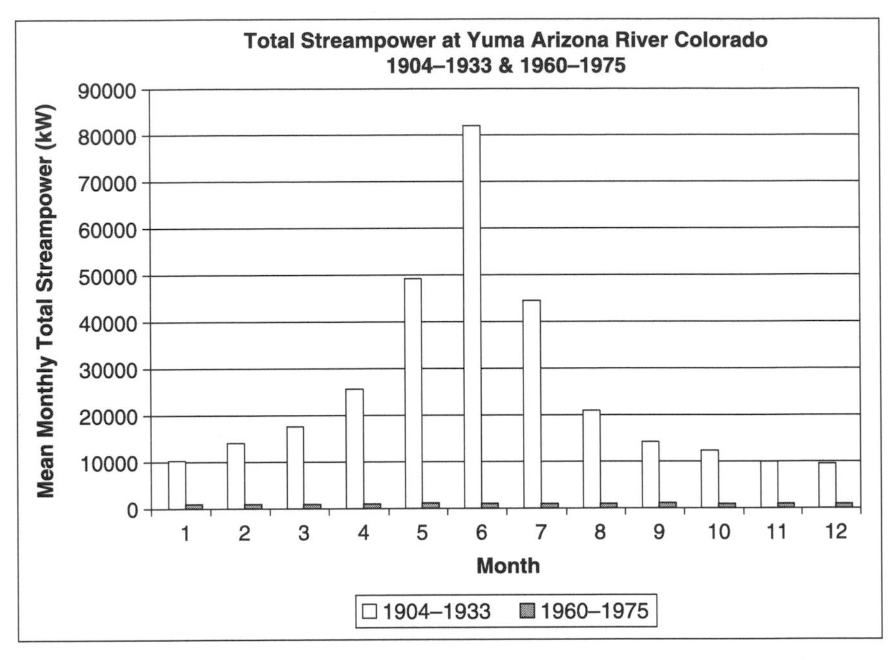

Figure 6.26 Total stream power at Yuma Arizona, Colorado River 1904–33 and 1960–75.

The conclusion reached was that there was probably no relationship between land use and the head height as a crude measure of power flux density. No relationship was indicated between land use and TSP for the reservoir reach. There was also unlikely to be a relationship between land use and former mean gradient. However, the results suggest there could be an inverse relationship between land use and the CSP (i.e. with slope and flow rate), although due to the small sample size the nature of the relationship could not be stated with confidence. A linear relationship between land use and water residence time could be indicated by the sample data, though again this could not be stated with much confidence. A linear relationship between land use and reservoir volume could be indicated, suggesting that despite differences in reservoir shape and depth, generally reservoirs of greater volume occupied more land per unit of energy generated. This implied that variations in reservoir shape and depth can be largely discounted as a factor accounting for the data, though again, due to the small sample size confidence in this was low. No correlation between land use and average reservoir fluid velocity was indicated. No clear correlation between land use and the production factor was apparent, though dams designed for considerable storage appeared to be the lower production factor schemes. No apparent correlation between land use and mean river flow was shown by the data.

Further interpretations and explanations of these relationships could be made; for example, the range in reservoir shape and the small size of the sample.

As stated in Chapter 3, land use per TWh might be expected to be inversely related to power flux density, in that the higher the power flux, the lower the area required. For HEP schemes this is mainly the reservoir area, which is a function of head height, flow rate, river gradient and the topography of the valley, as well as the storage capacity required. In the test sample, the greatest land area values were those of the desert or tropical HEP schemes, where low or relatively low flow rates coupled with highly seasonal flow variations are found. The Kariba scheme had the highest land use at 687 km^2/TWh per annum. This was followed by the Aswan High scheme at 500 km^2/TWh per annum. The Glen Canyon scheme followed with only 185.6 km^2/TWh p.a. and then the Hoover dam with 159.75 km^2/TWh p.a. The average land use for the sample was 175 km^2/TWh p.a.

These are all high-head impoundment HEP schemes, and most have low former river bed gradients in the range of 0.014 per cent – for the Aswan High dam – to 0.0519 per cent – for Glen Canyon – though the Hoover has a steeper gradient of 0.128 per cent. While the Aswan High had the lowest river bed gradient in the sample, there were other schemes with low former river bed gradients too, but it is apparently the combination of high-head schemes and low or relatively low bed gradients with relatively low flow rates that results in reservoirs with large land area per unit energy generation, in this selection.

In the sample the variation in valley topography, mean breadth and steepness of the valley sides may not be very conducive to the formulation of any reliable meaningful generalisation, for example relating the land use per unit energy to the head height, flow rate and bed gradient. In general it might be expected that the shallower the gradient and the broader the valley, together with low flow rates, the greater the land use per unit energy, and this is weakly indicated as a factor for this sample (see above).

However, the lowest land use per TWh per annum is that of the Three Gorges dam, at 12.8 km^2/TWh p.a., although this development has a low river bed gradient. The high flow rate is the crucial factor here. As pointed out earlier in this chapter, the steep sides and relatively low reservoir volume compared to the flow rate make for relatively efficient use of land area.

Land use: sedimentation correlation

There might appear to be a correlation between land use and sediment trap efficiency in the sample, with the Kariba dam using most land per unit of energy and also having the highest Brune trap efficiency. In fact, the five highest land use cases all have sediment trap efficiencies of over 96 per cent. The sample average Brune sediment trap efficiency was 76 per cent. The lowest land use cases in the sample had Brune trap efficiencies of 90 per cent or below, though the correlation was less pronounced in some cases. However, comparing

Brune sediment trap efficiency to land area km^2/TWh produced no clear correlation since the lower land use group of schemes had the greatest variation in trap efficiency from 18.1 per cent to 87 per cent, and the higher land use schemes all vary considerably from 159 to 687 km^2/TWh p.a.

6.3.1 Power flux density change in flow rates and head heights

Most impact might be expected where the change in river gradients of the former river bed to that of the reservoir, which is level, is greatest. Water residence times (i.e. the reservoir volume divided by the flow rate in the reservoir) may then relate to the greatest impacts.

The Aswan High scheme has the lowest former river gradient and also the second longest water residence time, with low average fluid flow rates and high sedimentation rates, with 98 per cent of 127 million tonnes per annum trapped – 125 million tonnes p.a. have been estimated (Stanley 1996; Saad 2002; El-Moattassem 1994). The change in power flux density in the large volume reservoir, together with long water residence time, causes a drop in water velocity, and allows deposition to occur.

The Brune sediment trap efficiency compared to head height might appear to suggest this for the sample (Figure 6.27). Were a trend line to be added, a linear correlation with a slope of $y = 0.774x$ and $R^2 = 0.5048$ would be shown, and a power correlation with a slope of $y = 2.9823x^{0.7264}$ and $R^2 = 0.8306$ can be shown. This could appear to accord with the hypothesis, in terms of hypothetical power flux density as resulting from head height. The river Nile might be expected to have a high sediment load, as it flows down from highland areas with high erosion rates, and flows through dry desert areas. The Three Gorges scheme has the second lowest former river gradient; again sedimentation might be expected. The scheme is expected to have high sedimentation rates of 541.16 million tonnes per year (Freer 2001), and 526 million tonnes p.a. at Gezhouba (Yishi 1994). However, this scheme, despite having an average sediment load cited as 1.2 kg per m^3 (Freer 2001), has a high flow rate, relatively low reservoir volume and thus relatively fast reservoir fluid velocity (average 0.218 m s^{-1}), reflected in the low water residence time, which counters to some extent the rate of deposition. Both of these rivers have high river loads, i.e. 127 million tonnes p.a. and 541.16 million tonnes p.a., respectively.

The magnitude of head heights may be related to the change in water velocity, the gradient over the former river bed being the operative factor here; the shallower the former river bed gradient, the greater the water velocity rate change per metre of head height. However, water residence times, depending on reservoir volume relative to flows, may appear more significant to the parameters of land use and sedimentation. High head height HEP schemes have other impacts, such as water quality, gas super saturation and fauna barrier effects, parameters that were outside the scope of the study.

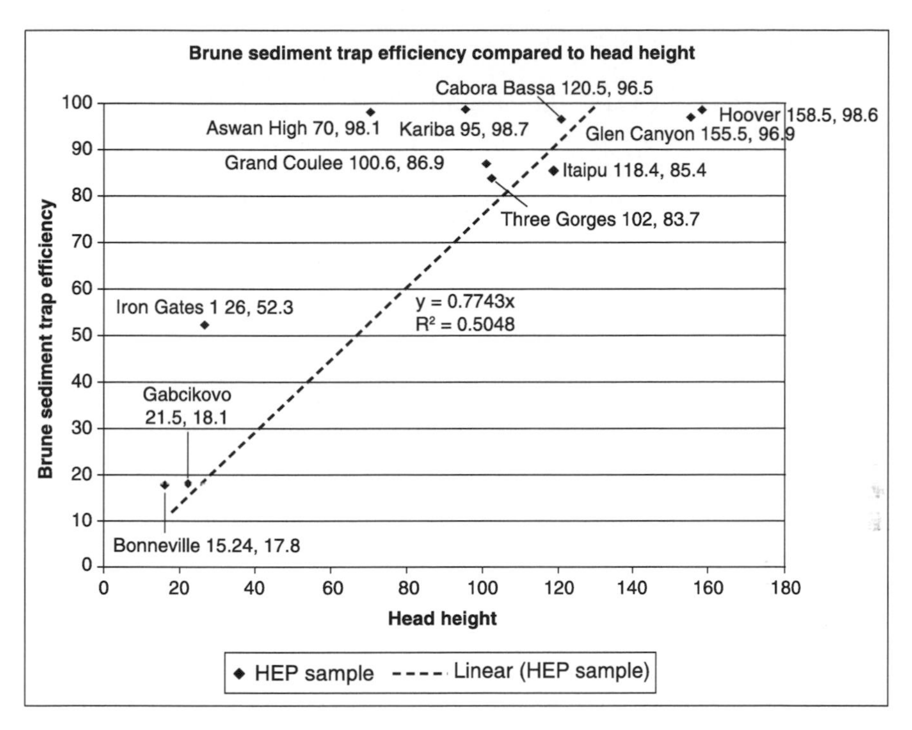

Figure 6.27 Brune sediment trap efficiency compared to head height and hypothetical power flux density as a reflection of head height.

6.3.2 Proportion parameter

The extraction and conversion of energy from the river power could be expected to result in impacts from loss of mechanical energy for erosion, and sediment load transport. The higher the proportion or ratio of energy extraction, the greater the losses expected. There are a number of different ways the parameter of proportion of flow can be measured:

- proportion of potential energy at the dam converted into electricity;
- proportion of the river flow abstracted via penstock and turbines;
- proportion of the river flow intercepted and converted to reservoir flow.

In terms of the conversion of river power to electricity, the Iron Gates I and the Itaipu schemes had high energy extraction rates of ~88 per cent and ~82 per cent, respectively, and therefore high mechanical energy losses from the river flow, which could lead to an expectation of losses of erosion and load transport, resulting in deposition. Deposition rates of the Iron Gates scheme have been cited as 20 million tonnes p.a. (Schwarz 2008) and for the Itaipu as 45 million tonnes p.a. (Borghetti and Perez 1994).

However, although the proportion of the river's power that is used for electricity produced can be cited, in terms of the functions of the river these functions are of necessity almost wholly interrupted when the whole of the river flow is impounded in the reservoir. The changes effected to the flow rates as the river enters the reservoir exhibit the change in the dissipation of energy from the equilibrium of gravity-induced kinetic velocity with bed friction, to near static conditions of potential energy for the larger reservoirs. Since the proportion of interception of the flow is 100 per cent (the dam straddles the whole valley), the whole of the river flow is affected. The degree of effect depends on the change in velocity and flow rates, and can also be expressed as the reservoir retention time.

The operation of normal impoundment HEP schemes involves varying reservoir levels. This may be done for a variety of purposes, for example to even out seasonal variations, to 'load follow' responding to demand fluctuations, to pass through flood flows or to reduce stresses on the dam. The consequence of lowering reservoir levels or 'drawing down' will be to reduce its depth and shorten the effective length of the reservoir, thereby reducing its volume. This can have the effect of increasing the water velocity and decreasing the retention time, depending on water flow rate. Therefore there can be less change in the water velocity than with full reservoir water levels. The lowered head height will mean less power can be converted to electricity. The power flux density at the dam will be reduced, but the power flux density over the reservoir reach will be increased.

The lowering of the reservoir levels in flood conditions is a tactic that has been used on a variety of dams for flushing through sediments (Tesaker and Tevinereim 1986; Yishi 1994), and is apparently to be employed on the Three Gorges HEP scheme in order to avoid excessive sedimentation. The ability to carry out such flushing depends on building bottom sluice gates into the dam design, though these are absent in older dams. As water velocity is a function of reservoir volume and river flow, the reduction in reservoir levels will, effectively, allow more of the sediment to be passed through, due to the higher water velocity in the reduced volume and shortened reservoir length.

This mode of operation can be described as reducing the proportion of energy abstracted from the river flow, over the reservoir reach, and increasing the proportion available for the river's functions, in this case that of transporting sediment.

Proportion: case of Gabcikovo

The change in flow rates and reservoir water retention time are useful indicators for potential impacts. However, this may not be the case for certain HEP schemes that have a reservoir 'offline', i.e. smaller schemes. The Gabcikovo scheme, for example, does not impound all of the flow for the full length of the diversion, but just for 16 km of the main

Hrusov–Dunakilit reservoir; the extension of a 17 km long leat, plus some further kilometres of tailrace canal below the power plant, takes about 85 per cent of the flow (Zinke 2004: 1; Liska 1993: 37). This was achieved after more water was allowed to flow in the old river course as a result of Hungarian government demands (Zinke 2004).

In the case of the Gabcikovo scheme, much of the impact is caused by the abstraction of ~85 per cent of the flow in the sealed raised leat for 17 km, leaving insufficient water flow in the side arms and in the old river bed to move sediments and flush through the gravel beds of the wetlands of this 'inland delta' section of the river (Zinke 2004).

These effects can be described as both losses of energy and power, i.e. TSP, to the reach concerned, and also as hydrology losses; for here the effect has been to lower the water table.

The effect of the unique (for its size) design of the Gabcikovo scheme is to reduce the land area use (fifth highest in the sample at 20km^2/TWh p.a.) and the reservoir volume in relation to the head height of up to 24 m achieved, but at the cost of diverting most of the flow away from the old bed, drying out the old wetlands. Originally, in the 1977 Treaty between Czechoslovakia and Hungary, the 'ecological' or compensatory flow was agreed as 50 m^3 s^{-1} in winter and up to 200 m^3 s^{-1} in the growing season (Liska 1993: 37). This would have provided compensatory flows of only 3.2–10 per cent of the seasonal mean. Zinke reported that as a result of (HEP) operations, parts of the extended system of braided channels and side arms had fallen dry, and by 2003 artificial methods of increasing water levels in these channels were being used. Small weirs to raise water levels were being used. In order to restore the wetlands of this area, Zinke has proposed that only 35 per cent of the flow should be diverted into the power canal channel, the leat, and that 65 per cent should flow in the river bed rather than the 15 per cent (Zinke 2004).

Proportion: the amount of energy that can be extracted

In seeking to discover how much energy can be extracted, while avoiding significant impacts, one response that can be suggested is: the proportion that leaves enough energy to perform the most critical functions of the river flow. What this could mean in practice can be seen from the examples cited in Chapter 5 and this chapter concerning the flood flow pulse releases on the Colorado River (USGS 1996), expressly for the purpose of ecosystem functions, such as sediment removal to preserve vegetation types and to reduce salinity and transport nutrients at the delta area.

The skewed distribution of TSP of the pre-dam flow, into the three months August, September and October, can be noted by considering the chart shown in Figure 6.28, of Aswan mean monthly TSP 1871–1967 and 1970–84. Sixty-two per cent of the total flow occurs in these three months and therefore a corresponding proportion of the river's energy. After dam construction, the flow rate reflects releases for the purposes of irrigation, water supply and power generation, with only a mild skew of 34 per cent of the total flow occurring in the three peak months of June, July and August.

Were some of the river's energy to be allocated to sediment transport and ecological functions in a similar manner to the Colorado River, pulse releases could be made, e.g. in the former peak flow month of September. Twenty-five per cent of the total river flow occurred in September prior to the dam, while this was reduced to just 7 per cent afterwards. A pulse release reflecting a proportion of the inflow to Lake Nasser/Nubia could, however, incur the loss of water stored for the winter and spring low flow months, particularly of April and May.

The pulse release power should be sufficient to transport sediment to replenish that lost by coastal erosion from the delta, and to counteract salinity incursions. These two impacts might be considered the most critical for the River Nile delta area. However, the stored Nile water allocations are very tightly committed for irrigation, water supply and energy production, and there is generally a water deficit in much of Egypt and the delta area (Stanley 1996). The Aswan High dam scheme was not designed to pass water allocations downstream for river functions and ecological purposes, and it is fair to say that environmental impacts were not fully anticipated. Egypt's population has grown, as have industrial activities, and water supplies are ever more stretched,

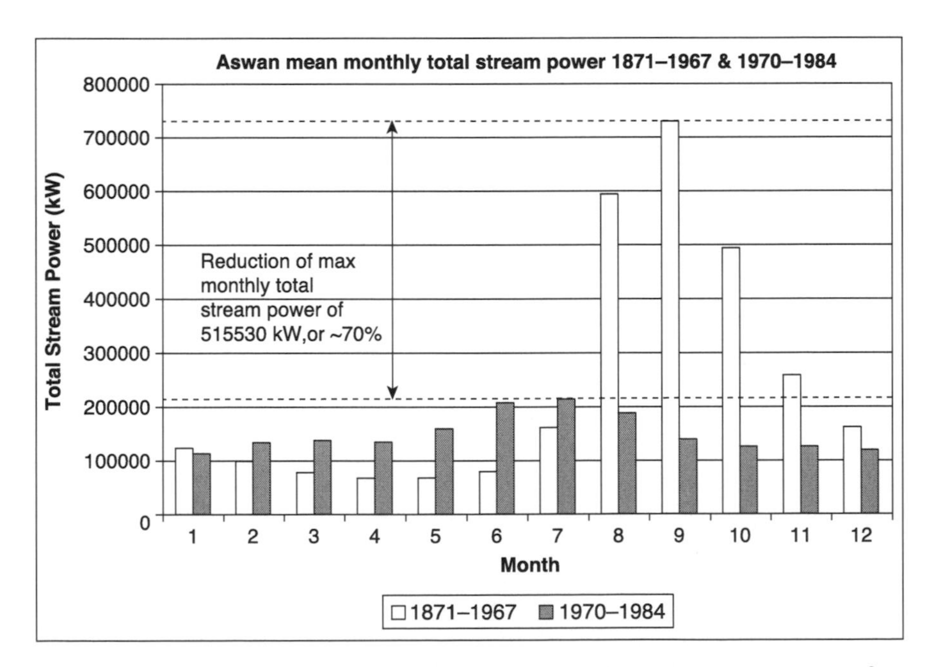

Figure 6.28 Reduction of maximum monthly mean total stream power at Aswan after construction of the Aswan High dam, by ~70 per cent.

making it hard to implement pulse releases. Whether retrofit modifications could be made to the Aswan High dam, e.g. bottom sluice gates, variable-height water intake and modification of irrigation supply downstream, so that the river could fulfil some of its natural functions, is a question of engineering, economic, political and ecological feasibility, though this could be desirable from an environmental perspective.

On the Colorado River, just 0.5 per cent of the total run-off was considered necessary to allocate to the river's functions at the Delta (IBWC 2001: 41) However, the main impacts being addressed were those of encroaching salinity, and some nutrient transport, as opposed to sediment transport, which would require greater flows.

The Aswan High scheme, which has the lowest gradient and the second longest water residence time in the sample, has a relatively high rate of siltation, with 98 per cent of the river load of 125 million tonnes p.a. (Stanley 1996; Saad 2002) deposited (El-Moattassem 1994).

The sample size of HEP schemes in this study has of necessity been small at eleven main cases and a further seven investigated in part, and hence cannot be statistically representative in relation to proportion of energy extracted and intercepted, but only indicative.

6.3.3 Total stream power analysis

The test of changes in TSP before and after dam construction, carried out on four of the rivers in the sample downstream of the dam, appears to indicate some relationship between power flux density and environmental impact. The reduction of the peak flows causes reduced TSP for those months, and thus the high power fluxes from floods are lost. Since most of the sediment transport and lifting of material to be transported occurs then, this function of the river flow is diminished or lost. The flood flows continue down the river, often all the way to the mouth and the sea. As a result, the loss of this power and energy extraction does indeed have consequences all the way down the river on sediment transport. Sediment transport is much reduced or has almost ceased.

TSP lost through attenuation of flood flows continues on downstream so that energy required for pick up and transport of material, suspended load and bed load is not available. The effect of this loss of TSP can be seen in the loss of material to replenish deltas, good examples of which are those on the Nile and the Colorado River.

Nile

The Nile mean monthly peak flows have been reduced by over 70 per cent at Aswan by the Aswan High dam scheme, with corresponding mean monthly TSP losses in the reach from Aswan to Gaafra, so that only 29 per cent of TSP is available below the dam. This diminution continues all the way down the Nile to the mouth. Formerly flood flows continued all the way to the sea before the Aswan High dam was built, transporting sediment and depositing it at the delta (Petts 1984). As a consequence mainly of this, the Nile delta is being eroded, with 100 metres per year retreat of the shoreline in some areas at Rosetta and up to 50 metres p.a. retreat in others, at Damietta (Stanley 1996). However, continuous data records covering the period before and after the dam were only available at Aswan.

Additionally, the maximum peak flows before the dam was built have been attenuated; 12,345 $m^3 s^{-1}$ is the maximum recorded, for the period 1871–1966 at Aswan, to a maximum of 2,639 $m^3 s^{-1}$ since for the period 1970–84. This represents a reduction of 78.62 per cent. The corresponding TSP values are 1,030,000 kW pre-1967 and 220,000 kW since, between 1970 and 1984 (see Figure 6.28).

Since it is the highest flood flows which are responsible for the highest TSP values, these peak values are responsible for the maximum sediment transport rates and the bulk of pick-up and transport. The highest flood flows perform most sediment transport and are inordinately important in, for example, recharging deltas (Stanley 1996).

The capability of the Aswan High dam reservoir, Lake Nasser and Nubia, to attenuate both flood flows and low flows is considerable, since the mean water residence time is nearly two years (1.94 years) due to its large volume of 168,900 × 10^6 m^3, relative to its mean flow rate of 2,760 $m^3 s^{-1}$.

Danube

In contrast to the Nile, however, on the Danube the flood flows are only slightly affected by the HEP schemes due to the dams being the 'run of the river' type. Since the reservoirs' volumes are relatively small in relation to flow rates, they are unable to significantly attenuate floods, and the seasonal peak flows are passed on downstream with the effect that the TSP is relatively unaffected and some sediment transport continues downstream; while overall the delta is diminishing, there is still some accretion in the north-western part (Panin and Jipa 1999). Sediment transport to the delta has reduced to 30–40 per cent of former rates, subsequent to the Iron Gates I and II scheme (Schwarz 2008).

For the 128.5 km Novo Selo to Lom reach, 29 km below the Iron Gates II scheme, the reduction in TSP after 1970 for the peak flow months of April and May was 6.9 per cent and 10.4 per cent, respectively. For the next section from Lom to Jiul, the TSP reduction is 6.14 per cent for April and 8.14 per cent for May.

The Iron Gates I reservoir volume is 2,500 × 10^6 m^3, and as a proportion of the mean flow rate of 5,500 $m^3 s^{-1}$, the mean water residence time is 5.26 days, which would be likely to result in a certain amount of storage time and thus some attenuation of peak flows, but not very much.

Much further down the Danube, at Zimnicea, 336 km below the Iron Gates HEP scheme, flood flows on the Danube after 1970 have been reduced by 22 per cent for April and 43 per cent for May. This greater reduction in TSP further downstream is possibly the cumulative result of multiple small dams and reservoirs retaining flows on tributary rivers entering the main stream (Schwarz 2008), as well as upstream. Schwarz cites loss of bed load as one of HEP's main consequences on the Danube. Also, the reduction in sediment transport is indicated by the Brune sediment trap efficiency estimates, of Iron Gates at 52 per cent, and Gabcikovo at 18 per cent.

At Dunaalmas, 87 km below the Gabcikovo HEP scheme, no loss of flood flow effect can be deduced from the data available. This is due first to the data set only extending to three years after the scheme began operating, and second it may well be that there is little or no attenuation occurring due to the scheme. This may be the case since the reservoir volume is small, at 243×10^6 m^3, as a proportion of mean flow rates of 2,047 m^3 s^{-1}, and mean water residence time in the reservoir at Gabcikovo is only some 32.97 hours. Therefore, flood flows could not be stored.

In this case, the flood flow rates and hence the TSP available still allows some material to be transported to the delta, though not enough to prevent the net balance of sea erosion and river deposition from causing diminution of the delta (Panin and Jipa 1999). However, diminution of the delta through erosion is also occurring due to channel dredging and jetty protection for improving shipping channels, as well as sea-level rise of ~3 mm p.a. (Schwarz 2008). Whereas Panin and Jipa put the blame on the Iron Gates dam project, an EU research project – EROS 2000 – states that studies show that 'most of the river's sediment is trapped by the delta' (EROS 2000). In contrast, according to the INHGA (2005), as much as a 65–80 per cent drop in sediment transport regime has occurred since the 1930–65 'natural river' period.

Columbia River

On the Columbia River, in the USA section below the Canadian border, after 1973 mean monthly peak flows reduced from 8,175 m^3 s^{-1} to 3,630 m^3 s^{-1}, i.e. to 44.4 per cent, at the Grand Coulee dam.

Assuming no dam had been built at Chief Joseph, the diminution of the peak TSP values would have been from 4,090,000 kW to 1,820,000 kW in June for the 82.1 km reach, with a fall of 51 m, and can be observed from Figure 6.29. This can also be seen in Table 6.5, which includes both the actual level gradient of the current Lake Rufus, with TSP values of zero kW, and the values assuming the original surface gradient profile. The mean TSP values do correlate roughly with the mean power flux of the river at the dam line, based on the formula 'Power in kW = $q \times g \times h$' obtained from other sources, allowing for differences in head height due to varying reservoir levels. The power flux is 1,630,000 kW, based on a head height of 54.25 m

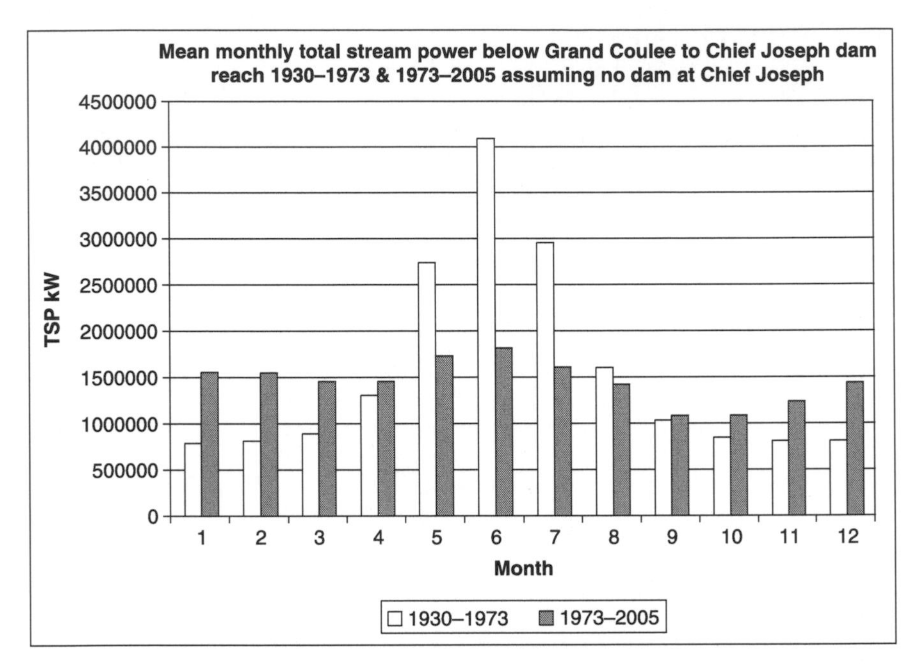

Figure 6.29 Mean monthly total stream power below Grand Coulee to Chief Joseph reach 1930–73 and 1973–2005, assuming no dam at Chief Joseph, kilometre 877.

for Chief Joseph dam, while the mean TSP values based on 51 m height difference comes to 1,560,000 kW.

It can be seen that construction of Chief Joseph HEP scheme has resulted in an almost complete loss of TSP over the reach, with the river power now concentrated at the dam. Formerly, this energy was dissipated by friction with the river bed and in the transport and suspension of sediment. Since the US section of the river is almost entirely impounded with dams, the TSP of the river has correspondingly been almost completely lost and the river power is now mostly concentrated at each of the dams, where it is used to generate electricity, or passed over the dam spillways.

In the only natural reach left on the US stretch of the Columbia River, that of the 95.12 km reach from below Priest Rapids dam to the headwaters of McNary reservoir, the maximum mean monthly TSP has fallen from 2,050,000 kW in June between 1930 and 1972, to 1,010,000 kW between 1973 and 2006, i.e. 49.42 per cent, as a result of the impoundments upstream, especially those in Canada, the Mica dam particularly. This reduction in peak flows and thus TSP can be observed all the way downstream to the mouth.

In addition to the loss of TSP and its effects on sediment transport, one of the major impacts of the Columbia River is the reduction in steelhead salmon fisheries due to the impoundments (Petts 1984). The change in habitat from

Table 6.5 Mean monthly total stream power values in kilowatts for the Grand Coulee to Chief Joseph reach of the Columbia River 1930–73 and 1973–2005, assuming original surface gradient profile and actual surface gradient

Month	*1930–73 (kW)*	*1973–2005 assuming original gradient profile (kW)*	*1973–2005 actual gradient*
January	789,000	1,560,000	0
February	813,000	1,550,000	0
March	892,000	1,460,000	0
April	1,310,000	1,460,000	0
May	2,740,000	1,730,000	0
June	4,090,000	1,820,000	0
July	2,950,000	1,610,000	0
August	1,602,000	14,200,004	0
September	1,030,000	1,080,000	0
October	846,000	1,090,000	0
November	809,000	1,240,000	0
December	811,000	1,440,000	0

free-flowing river with sections of rapids and waterfalls and with gravel beds, to a series of lakes, has been profound and has a variety of consequences. One of these, the reduction of seasonal flow variations, is connected to the poor breeding rates of surviving salmon, which depend on seasonal flooding for salmonid movement downstream (Petts 1984). The diminution of the mean seasonal flow rate range (i.e. the difference between peak and low flows) from 6,597 $m^3 s^{-1}$ to 1,465 $m^3 s^{-1}$, just 22.21 per cent, is marked.

Colorado River

Although the Colorado has been much less impounded than the Columbia and still has considerable natural gradient reaches, its two major dams have a large volume compared to the mean flow rate, and hence long water residence times. For the Glen Canyon scheme, the residence time is 357 days, almost a year (0.9785 years), and for the Hoover scheme it is 1,024 days, almost three years (2.8076 years). The highly variable seasonal flow rate of the natural river is significantly attenuated, and this is partly the purpose of these schemes; that is, to ensure a reliable flow for the multiple (human) purposes.

At Lees Ferry 23.4 km below Glen Canyon dam, the mean monthly peak TSP in the period 1911–63, of 1,830,000 kW fell to 528,000 kW for the reach Lees Ferry to the Little Colorado confluence. This was 28.85 per cent of the former TSP, or a reduction of 71.14 per cent. This can be seen from Figure 6.30, together with the diminution of the monthly seasonal variation from 1,660,000 kW to 1,840,000 kW – just 11.07 per cent, or a reduction of 88.93 per cent.

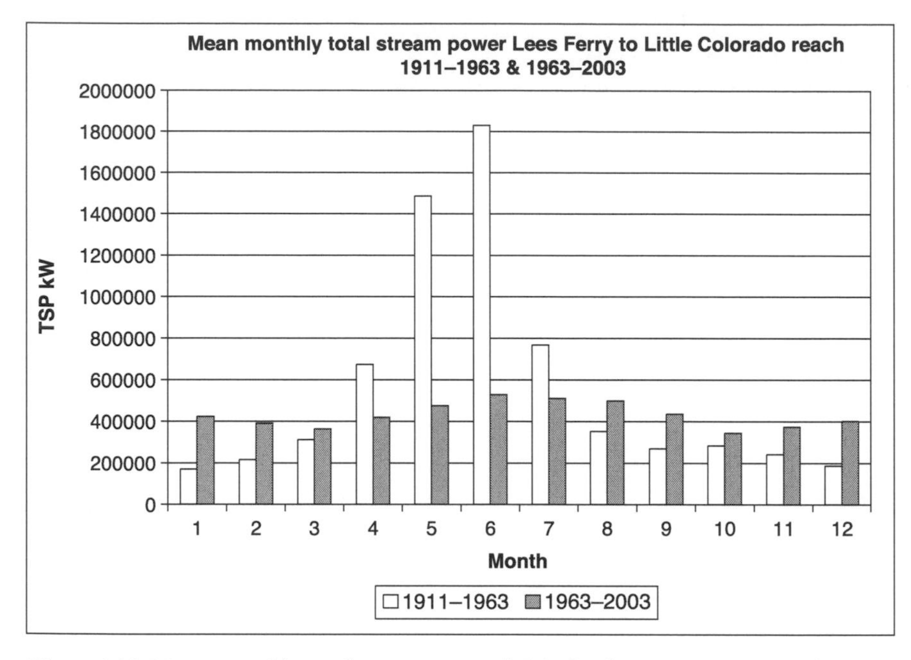

Figure 6.30 Mean monthly total stream power (kW), for the Lees Ferry to Little Colorado reach, Colorado River, 1911–63 and 1963–2003.

Below the Hoover dam, similar reductions in peak flows and thus TSP can be observed after 1933 (Figure 6.31), an effect which is clearly apparent some 500 km lower downstream at Yuma, and then after 1958–64, the building of the Glen Canyon dam adds to the effect. However, the main factor influencing TSP at Yuma is the abstraction of water for irrigation and domestic, urban and industrial water supply, shown by the diminution of mean flow rates at Yuma from 613 $m^3 s^{-1}$ prior to 1933, to 187 $m^3 s^{-1}$ between 1934 and 1959, and then just 18 $m^3 s^{-1}$ post-1975.

Although this water abstraction has not been for the purpose of energy, but for water supply, it does impact greatly on the river's available energy to perform its functions. The mean TSP at Yuma Arizona had fallen by 96.18 per cent, with an even greater loss of 98.6 per cent TSP in the peak flow months, and represents an extreme example of the abstraction of a natural energy flow.

The resulting almost complete cessation of sediment entrainment and transport is reflected in the very considerable losses of the delta area at the mouth of the Colorado in Mexico (IBWC 2001), as well as river bed and channel effects.

6.3.4 Changes in TSP compared to delta land area loss

The changes in TSP before and after dam construction on three of the rivers in the sample downstream of the dam were compared with delta land loss, as a measure of impact from reduced sediment transport (see Table 6.6).

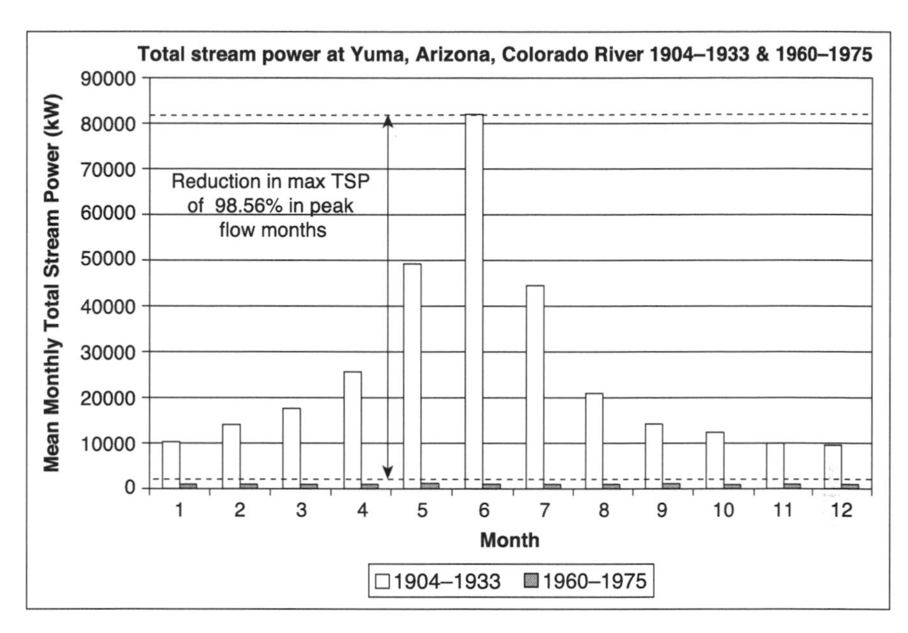

Figure 6.31 Reduction in maximum monthly mean total stream power at Yuma, Arizona of 98.56 per cent, 1904–33, as compared to 1960–75.

Peak TSP change compared to delta land area loss

Delta land area loss was ascertained from literature sources, with estimations used, for example, in the case of the Nile delta, where no clear figures for land loss were discovered. Land loss rates of up to 100 m per year at Rosetta and 50 m per year at the Damietta stream promontories have been cited (Stanley 1996). While overall land area is being lost, some locations are experiencing accretion due to long shore wave erosion and transport (Blodget *et al.* 1990). Therefore, perhaps partly as a result of the dynamic nature of deltas, the author has been unable to find a clear figure for land area loss in the literature. Additionally, other causes of land area loss are also cited, such as sea-level rise (Frihy and Khafagy 1991), blocking of drainage channels and subsidence due to freshwater well pumping and drilling for oil and gas (Frihy and Khafagy 1991; Stanley and Wingerath 1996). However, reduced sediment transport by the river is undoubtedly the major factor, as until about 1900 accretion of the promontories by 3–4 km over the period between 1800 and 1900 was recorded (Lofty and Frihy 1993). An estimation of land area loss has been made based on these diverse sources. The length of the coastline of the Nile delta has been cited as 225 km (Stanley 1996), and maximum coastal retreat of 3–4 km since 1900 has been recorded. If an average retreat of 0.5–1 km is assumed, then ~100–225 km^2 can conservatively be assumed to have been lost since dams were first built at Aswan. This land

Table 6.6 Peak total stream power change compared to estimated approximate delta land area loss

River	*HEP scheme year*	*Peak stream power change (%)*	*Estimated delta land area change*	*Estimated land area loss per year*
Nile	Aswan High 1967 and (Old Aswan 1904)	–71	112 km^2 to 225 km^2, or 0.5–1% loss since 1900	1.12 km^2 –2.25km^2
Danube	Iron Gates I 1970 and Iron Gates II	–6.9 –3.36	19.6 km^2 or 0.24% loss since 1975	0.65 km^2
Colorado	Hoover 1933 Glen Canyon 1963 Yuma	–71.14 –98.6	1,874 km^2, or 76% loss since 1900	18.74 km^2

area loss since 1900 has been turned into an estimated figure of 1.12–2.25 km^2 lost per year. However, the major loss has occurred more recently since the Aswan High dam and is probably higher, so this figure is considered conservative.

For the Danube delta, over the period 1975–2003, about 100 ha per year – over 100 km of the Black Sea coastline – has been lost as a result of intensive erosion processes, with 70 km of this coastline within the delta area (Zoran and Anderson 2006). Therefore, assuming 28 yr × 70 ha = 1,960 ha, or about 19.6 km^2 has been lost, which equates approximately to 0.65 km^2 p.a.

The Colorado delta has been cited as having an area of 24 per cent of that it covered in the early 1900s (Hinojosa-Huerta 2004). The land area loss by this reckoning has thus been 1,874 km^2, or ~18.74 km^2 p.a.

Figure 6.32 shows the reduction in peak stream power compared to delta land area loss over periods since impoundment and water management and diversion for three of the rivers in this study which have deltas. Were a trend line to be fitted, a linear slope of $y = 13.564x$, with $R^2 = 0.5945$, could be shown, but is not significant, also due to the very small sample. The land area loss for the Nile delta is estimated at 225 km^2, but may be considerably higher. A correlation between peak TSP and delta land area loss is indicated, but with only three data points these figures cannot convey very much certainty. For rate of land area loss per year, see Figure 6.33. If a trend line were to be included, a linear slope of $y = 0.1358x$, with $R^2 = 0.5845$ could be shown, i.e. not significant, and as the sample is so small, it is not included. If the lower range of the Nile delta land area loss per year is taken, a similar slope would be shown.

The peak TSP reduction has been compared with Brune sediment trap efficiency in Figure 6.34 and appears to indicate a linear correlation; were a trend line drawn, the slope would be $y = 0.7173x + 48.39$, with $R^2 = 0.9911$. However, there are too few data points for significance.

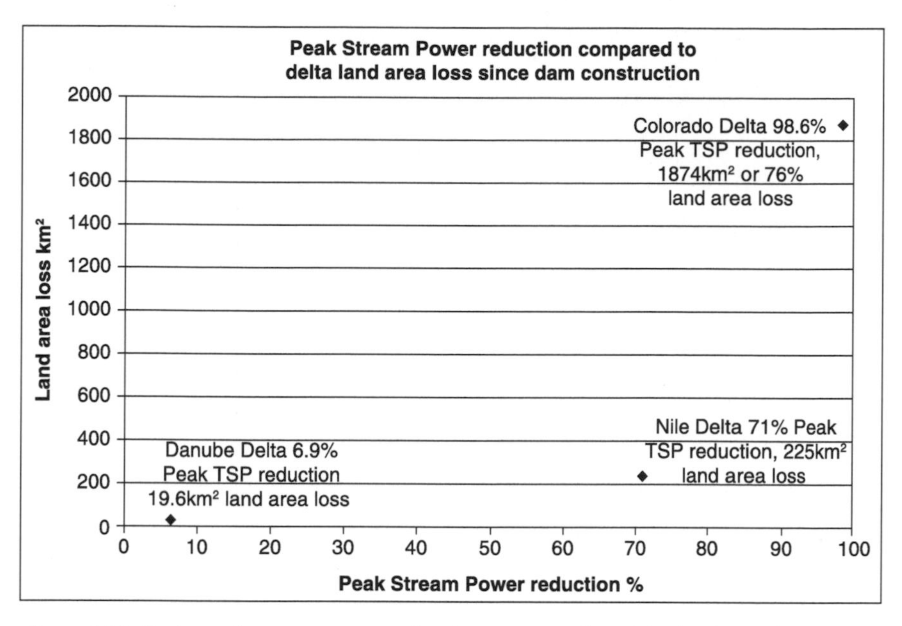

Figure 6.32 Peak total stream power reduction compared to delta land area loss since major dam construction.

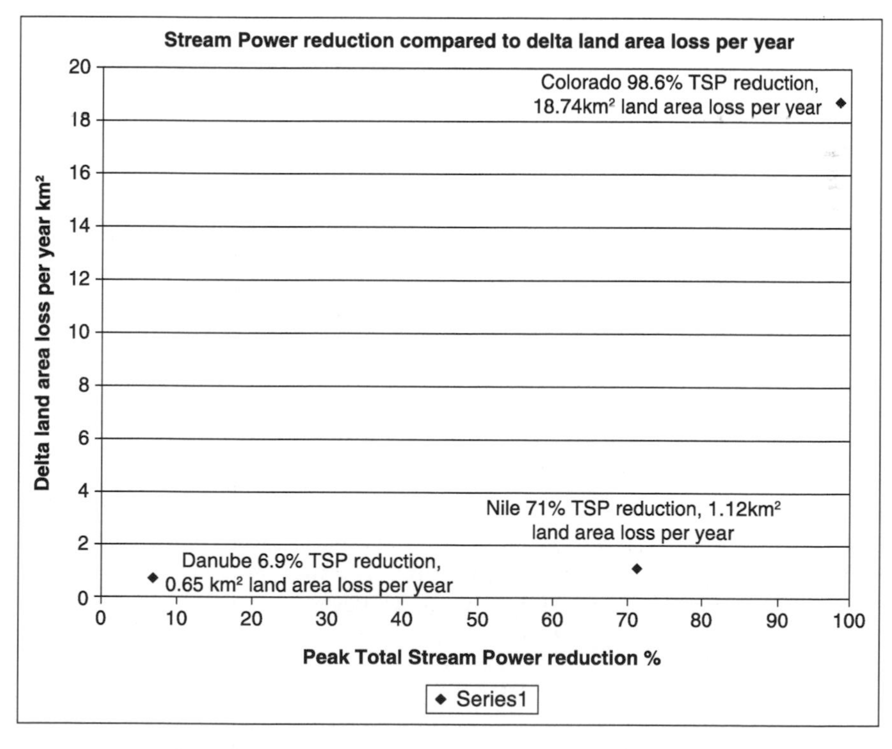

Figure 6.33 Peak total stream power reduction compared to delta area land loss per year.

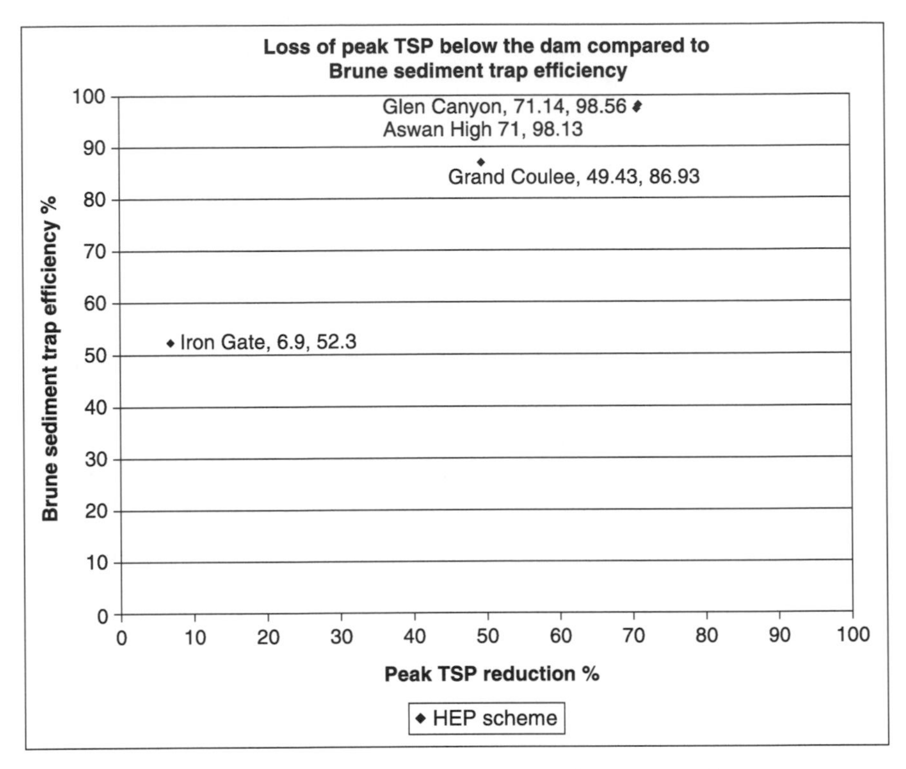

Figure 6.34 Loss of peak TSP below the dam compared to Brune sediment trap efficiency.

For comparison, the delta area land loss per year has been compared with Brune sediment trap efficiency (Figure 6.35). The comparison of Brune sediment trap efficiency with delta land area loss shows some indication of a correlation, though given the small sample of five, and the uncertainty over the Nile delta land loss area, it is not significant.

One of the conclusions suggested is that the development of a large number of smaller dams with small reservoir volumes could result in lower changes to peak TSP downstream, i.e. less change in power flux density, with correspondingly less change to sedimentation, as well as sediment transport, than a few large impoundment dams. In terms of sedimentation in the reservoir reach, and downstream TSP, the perpetuation of seasonal peak flows and thus sediment transport, fewer impacts would be experienced. The changes to the river bed would be reduced and deltas could still be recharged. Such HEP schemes, being of the 'run of the river' type, would result in an overall reduction in electricity production, as the peak flows would be passed over sills or through sluices rather than being stored and released at a constant rate. The power of these peak flows would be available to the river to perform the functions outlined earlier, and available as stream power.

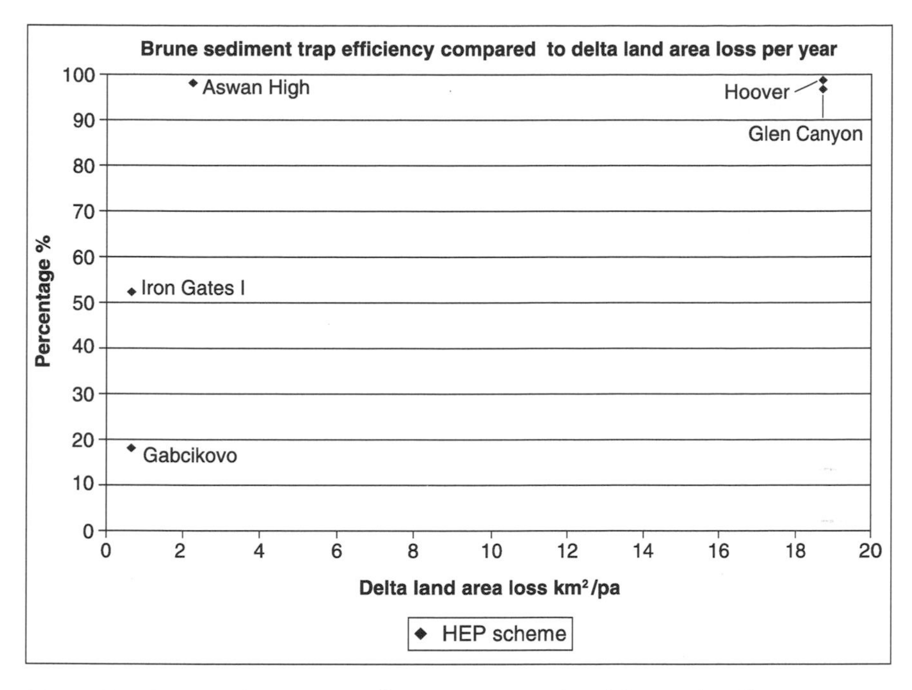

Figure 6.35 Brune sediment trap efficiency compared to delta area land loss per year.

This would represent a reduction in the proportion or ratio of energy abstracted from the river as a whole. As an example of this approach, the comparison between different developments of HEP on the four river systems can be cited. The Danube delta has lost some area – 0.24 per cent, and is losing 0.65 km^2 p.a. – partly as a result of HEP development on the river and its tributaries, while the deltas of the Nile and the Colorado are not being recharged sufficiently and are losing area at a greater rate, estimated at 1.12–2.25 km^2 and 18.74 km^2 p.a., respectively. The Danube's HEP schemes are 'run of the river' ones which do not affect the TSP very much, while the Nile and the Colorado have large impoundment dams which significantly reduce TSP. Conversely, the Danube's dams cannot hold back flooding from large seasonal peak flows, and there have been floods, such as in 2001 (JDS 2005) and 2013.

The Three Gorges scheme is interesting in this regard as it has been decided to operate it by lowering the reservoir level substantially during peak flows in order to prevent silting up (Jisheng 1994). The head height was to be reduced from the 102–105 m (~175 m above sea level) optimum to 71 m (145 m above sea level) (Jiazhu 1994) in the flood period (Xuezhong 1996: 24). The lowering of head heights in anticipation of flood flows is common practice worldwide, so that there is capacity to store flood waters, though this is more marked in the case of the Three Gorges. However, it would appear, in the author's view, that in this case if the water level were to be kept at 71 m height by sluice gate control, then its effectiveness in storing flood water would be reduced. This would effectively reduce the proportion of energy extraction from the river, and transfer it back to the river to result in increased water velocity in the reservoir (as a result of the reduced cross-section and volume), and hence more sediment transport. Were this to be so, then this could to some extent reduce its ability to hold back flood flows, which is one of the claimed purposes of the scheme. Nevertheless, the dam will, it is claimed, have the ability to reduce the 1870 peak flood flow (0.1 per cent flood frequency) of 110,000 $m^3 s^{-1}$ to below 80,000 $m^3 s^{-1}$ in the vulnerable Jingjiang stretch, as well as reducing 1 per cent flood frequency events to an acceptable less than 60,000 $m^3 s^{-1}$ flow there (Jiazhu 1994). The dam has apparently already proved its effectiveness in reducing flooding below the scheme in 2009 and 2010, with outflows reduced to 40,000 $m^3 s^{-1}$ according to the SINA Corporation. The avoidance of silting in the upper reaches is necessary if the advantages of improved navigation conditions are to be maintained at the port of Chongqing at the upper end of the reservoir (Jisheng 1994). At the same time the vital function of recharging of the delta at Shanghai could be better maintained.

Therefore, by reducing the power flux density at the dam (by reducing the head height), it would appear that the proportion of energy extracted can be reduced and some important impacts avoided. This latter is an important observation, which is considered to endorse the hypothesis.

Bibliography

Bagnold, R.A. (1966) 'An Approach to the Sediment Transport Problem from General Physics', USGS, Professional Paper.

Blodget, H.W., Taylor, P.T. and Roark, J.H. (1990) 'Shoreline Changes Along the Rosetta-Nile Promontory: Monitoring with Satellite Observations', *Marine Geology*, Vol. 99: 67–77.

Borghetti, R. and Perez, D. (1994) 'Socio-Environmental Actions on the Itaipu Binacional Reservoir', *International Water Power and Dam Construction*, Vol. 46 No. 7: 36–8.

El-Moattassem, M. (1994) 'Field Studies and Analysis of the High Aswan Dam Reservoir'. *Water Power and Dam Construction*, January.

EROS (2000) 'The Interactions Between the Danube and the North West Back Sea', Environment Institute, Joint Research Centre, Italy, online resources: http://ec.europa.eu/research/intco/pdf/094e.pdf, accessed 17 August 2009.

Freer, R. (2001) 'The Three Gorges Project on the Yangtze River in China', *Proceedings of ICE, Civil Engineering*, Vol. 144: 20–8.

Frihy, O.E. and Khafagy, A.A. (1991) 'Climate and Induced Changes to Shoreline Migration Trends at the Nile Delta Promontories, Egypt'. *CATENA*, Vol. 18 No. 2: 197–211.

GRDC (2008) Data from the Global Runoff Data Centre, Koblenz, Germany.

Hinojosa-Huerta, O. (2004) 'Restoring the Colorado Delta', Southwest Hydrology, online resources: www.swhydro.arizona.edu/archive/V3_N1/feature5.pdf, accessed 9 April 2010.

IBWC (2001) 'United States–Mexico Colorado River Delta Symposium 11–12 September 2001', online resources: www.ibwc.state.gov?FAO/CRDS0901/English Symposium.pdf, accessed December 2008.

INHGA (2005) 'Water Flow and Sediment Transport in the Lower Danube River', *Geophysical Research Abstracts*, Vol. 7, online resources: www.cosis.net/abstracts/EGU05/05967/EGU05-J-05967.pdf, accessed 17 August 2009.

JDS (2005) ' Joint Danube Survey 1', International Commission for the Protection of the Danube River, online resources: www.icpdr.org/icpdr-pages/jds.htm, accessed 22 July 2009.

Jiazhu, W. (1994) 'Major Problems with the Design of Three Gorges', *International Water Power and Dam Construction*, August: 24–32.

Jisheng, C. (1994) 'Sedimentation Studies at Three Gorges', *International Water Power and Dam Construction*, August: 54–8.

Liska, M.B. (1993) 'Gabcikovo–Nagymaros: A Review of Its Significance and Impacts', *Water Power and Dam Construction*, July.

Lofty, M.F. and Frihy, O.E. (1993) 'Sediment Balance in the Near Shore Zone of the Nile Delta Coast, Egypt', *Journal of Coastal Research*, Vol. 9: 654–62.

Marndouh, S. (1985) *Hydrology of the Nile Basin*, Elsevier, London, Amsterdam and New York.

Panin, N. and Jipa, D. (1999) 'Danube River Sediment Input and Its Interaction with the North West Black Sea', *Estuarine Shelf & Coastal Science*, Vol. 54: 551–62.

Petts, G.E. (1984) *Ecology of Impounded Rivers: Perspectives for Ecological Management*, John Wiley & Sons, Chichester.

Rzoska, J. (1978) *On the Nature of Rivers*, Dr Junk Publishers: The Hague, Boston, London.

Saad, M.B.A. (2002) 'Nile River Morphology Changes Due to the Construction of the High Aswan Dam in Egypt', 5th International Conference of ICHE, online resources: http://kfki.baw.de/fileadmin/conferences/ICHE/2002-Warsaw/ARTICLES/PDF/248C4-SD.pdf, accessed 24 July 2009.

Schwarz, U. (2008) 'Assessment of the Balance and Management of Sediments of the Danube', FLUVIUS Floodplain Ecology and River Basin Management, online resources: http://ksh.fgg.uni-lj.si/bled2008/cd_2008/05_Floods,%20morphological%20processes,%20erosion,%20sediment%20transport%20and%20sedimentation/211_Schwarz.pdf, accessed 23 July 2009.

Stanley, D.J. (1996) 'Nile Delta: Extreme Case of Sediment Entrapment on a Delta Plain and Consequent Coastal Land Loss', *Marine Geology*, Vol. 129: 189–95.

Stanley, D.J. and Wingerath, J.G. (1996) 'Nile Sediment Dispersal Altered by the Aswan High Dam: The Kaolinite Trace'. *Marine Geology*, Vol. 133: 1–9.

Tesaker, E. and Tevinereim, K. (1986) 'Hydro Electric Flushing of Sediments at Small Hydro Plants'. *International Water Power & Dam Construction, Second International Conference Proceedings*, Vol. 38: 365–71.

USGS (1996) 'Calculated Hydrographs for the Colorado River Downstream from Glen Canyon Dam During the Experimental Release March 22nd to April 8th 1996', US Department of the Interior, USGS, online resources: http://pubs.usgs.gov/fs/1996/fs-083-96/:www.ibwc.state.gov?FAO/CRDS0901/EnglishSymposium.pdf, accessed 17 June 2009.

Xuezhong, L. (1996) 'Accentuate the Positive', *International Water Power & Dam Construction*, November: 24–5.

Yishi, L. (1994) 'The Gezhouba Project in Operation', *International Water Power & Dam Construction*, August.

Zinke, A. (2004) 'Hydropower Station Gabcikovo: Deficits in Hydrology (Sediment Transport, Groundwater) and Biology', International IAD Workshop, 14–16 October, online resources: www.zinke.at/Zinke.data/Images/Zinke-IAD.Gabcikvo.122-04.pdf.
Zoran M. and Anderson, E. (2006) 'The Use of Multi-Temporal and Multispectral Satellite Data for Change Detection Analysis of the Romanian Black Sea Coastal Zone', *Journal of Optelectronics and Advanced Materials*, Vol. 8, No. 1: 252–6.

7 Extension of the hypothesis to water-based renewable energy sources

7.1 Introduction

Chapter 6 described the conclusions from the test of the hypothesis on selected cases of HEP developments and downstream rivers. Using a variety of parameters at the reservoir reach and the stream power concept for the river downstream, it was tentatively concluded that the greater the change in power flux density the greater the impact was likely to be. The other parameters of the hypothesis – proportion of energy extracted, efficiency of conversion and number of conversions – were proposed as related to impact, though not tested. In this chapter the hypothesis is extended to other water-based renewable energy flows.

Extending the hypothesis from HEP to other water-based renewable energy sources which have relatively dense power fluxes, the nearest comparable source to HEP is tidal barrage technology, being akin to low-head hydro plant (Elliott 1996). Related to tidal barrages are marine current turbines, which also harness tidal currents, but without a barrage, and so this technology is also considered in terms of the hypothesis. Although tides are often perceived as currents, they can be considered to be very low-frequency waves with very long wavelengths (MacKay 2007). Progressing from 'tidal' waves, wind-derived wave energy is the third technology considered here in terms of the hypothesis.

The parameters of the impact model proposed in this book are:

- the power flux density of proposed tidal and lagoon schemes;
- the proportion of the energy in the flow extracted;
- the number of conversions;
- the conversion efficiency;
- the changes to the existing flow patterns.

Here, these parameters are applied to the non-HEP water-based renewable energy technologies.

7.2 Tidal barrages

While tidal barrages may resemble hydro electric schemes, they differ in that energy is extracted from a very low-frequency oscillating wave motion, occurring approximately twice per 24 hours, although in certain locations only once (Baker 1988). Generation is performed by a low-head hydro plant, from a continuously varying head height obtained (usually) by the capture and retention of the flood tide in a reservoir, generation being on the ebb tide, though in some cases on the flood tide. As such, tidal barrages are both intermittent and highly variable sources of renewable energy, though completely predictable (Baker 1988). There are a number of different operating modes for a tidal barrage scheme. They can generate on the ebb tide only, or on the flood tide, or in a combination of these two modes. At the high tide, pumping can raise the basin head height; at low tide, pumping from the basin can increase the differential in the head heights to obtain more energy capture (Baker 1988). Furthermore, a multiple-basin arrangement can be designed in order to produce more continuous generation by timing the alternate filling and emptying to coincide with tides. However, experience so far has determined that the maximum energy can be generated by ebb-only operation (Baker 1988; Charlier 2003).

7.2.1 Tidal barrages around the world

There are as yet very few full-scale tidal barrage schemes anywhere in the world, with only two full-scale schemes operating at the time of writing – the Rance Barrage in France, with a 240 MW capacity dating from 1966 (Charlier 2003) and the slightly larger Korean Sihwa Lake (south coast) 254 MW tidal barrage plant commissioned in 2011. This is followed, in order of size, by the Annapolis tidal plant of 18 MW in the Bay of Fundy, east Canada, then the Janxia Creek 500 kW in the East China Sea, followed by the 400 kW Russian Kislogubsk plant and a number of micro plants in China (Charlier 2003). There have been numbers of schemes for tidal barrages planned around the world, as the technical potential has been assessed at 2.6 TW, about one-tenth of that of HEP (Baker 1988). Examples include: the Korean Sihwa Lake (south coast) 254 MW tidal barrage plant under construction since 2004, in an existing embankment (Ford 2006), now commissioned; the Korean 520 MW Garorim Bay; and the Chinese planned 300 MW tidal lagoon at the mouth of the Yalu River (Ford 2006). In the UK the existence of the second highest tidal range in the world at the Severn Estuary has led to numerous plans and studies for a barrage of 8.6 GW generating capacity. Studies have been carried out for other sites, such as the Mersey, the Solway Firth and Morecambe Bay, as well as numerous smaller estuaries (Sustainable Development Commission 2007).

Generally, the economics have not proved favourable yet for most of these schemes due to the lengthy major civil engineering work required for the long barrages, and the low and varying head (Ford 2006). However, as the successful operation of the Rance Barrage since 1966 demonstrates, as well as other smaller examples, the technology is proven and practical.

7.2.2 Environmental impacts of tidal barrages

In the UK, environmental considerations have also played a part in preventing construction of tidal barrages. The Severn Estuary scheme with ebb generation would result in continuous submersion of roughly half of the area of mudflats uncovered at low tide, which provides an important feeding ground for bird species (Shaw 1980). The area has the international designation of a RAMSAR site, under which a site is 'considered internationally important if it regularly supports 1% of the individuals in a population of one species or subspecies of waterbird' (Joint Nature Information Committee, 2008). Five species of bird have more than 1 per cent of their north-west or UK population supported at this site: shelduck (1.8 per cent), dunlin (2.7 per cent), curlew (1.0 per cent), Bewick's swan (5.0 per cent), widgeon (1 per cent of UK population) (Joint Nature Information Committee, 2008). The Severn Estuary is also designated as a European Special Protection Area under the EC Birds Directive (EC Birds Directive 1979).

While a reduced tidal range inside the barrage is a feature of tidal barrage schemes, the individual sites will vary considerably; some will involve long barrages across broad bays or inlets, others may have only short barrages across narrow mouths leading to long, and in some cases deep, estuaries. Hence the precise impacts will tend to be site-specific, for example the extent to which mud banks are immersed will depend on the shape of the basin, the depth of the basin and steepness of its banks, as well as variations in the tidal range.

A comprehensive review of impacts is provided by Clarke (1995), as well as elsewhere, and it is not proposed to repeat all of these here. Some of the main potential impacts likely to be expected are summarised in Table 7.1.

7.2.3 Power flux density of tidal barrage schemes

Tidal barrages have significantly lower power flux densities than most large HEP plants. This is due to the relatively low head heights and their variability over the tidal cycle. Considering power flux densities for tidal energy, this is shown in Figure 3.3 as having a ~35 kW/m^2 (Clarke 1993) maximum power flux density and an average of about 6 kW/m^2 (based on power fluxes in the estuary before entrainment by barrage structure, or capture). However, as described below, this may be too low in the case of the Severn Barrage.

Table 7.1 Summary of general impacts from tidal barrages

- Reduced tidal range – 50 per cent
- Loss of mud flats
- Tidal changes downstream
- Reduced/raised water speed
- Siltation
- Scouring
- Lacustrine environment created
- Changed salinity
- Changed turbidity
- Fish migration barrier
- Fish pressure damage
- Wildlife changes
- Drainage effects
- Possible water table effects
- Raised water level for shipping
- Flood protection
- Shipping hazard

Source: Clarke (1993).

According to Charlier (2003), the potential energy is given by the equation:

$$E_p = K \times 10^6 \times A \times R^2$$

where:
K = constants of 1.92 or 1.97
A = area of basin
R = average tidal range
(Bernstein 1961; Gibrat 1966; Mosonyi 1963; cited in Charlier 2003)

For the Severn Barrage the potential energy available assuming a basin area of 70 km^2 according to this method is:

$$E_p = 1.92 \times 10^6 \times 70 \times 9.8^2 \text{ (in kWh per year)}$$
$$E_p = 12.908 \times 10^{10} \text{ kWh p.a. or } 12.908 \text{ TWh p.a.}$$

This value of ~12.9 TWh p.a. is roughly equivalent with the 17 TWh of electricity per annum that it is estimated a barrage at the Severn Estuary would generate, given the assumptions of 70 km^2 basin area and a mean tidal range of 9.8 m (from Hubbert 1969; cited in Twidell and Weir 1986: 353). (Elsewhere the Severn Estuary barrage reservoir area is cited as 480 km^2 in area, which would result in 88.51 TWh) (Department of Energy 1989).

Charlier (2003) describes capacity factors as a

> type of utilization factor which is based on the system selected: for a single or double basin plant and a single tide it is 0.224, double tide 0.34

but 0.21 for a double basin; for a double basin, single tide with reverse pumping it is 0.277 and with pumps 0.234.

A capacity factor of about 0.224 accords roughly with the capacity factor quoted for the Severn Barrage scheme. This value, a little under one-quarter, relates the average power output to the peak power output capacity.

Another method of calculating the average potential energy available is the equation:

$$\text{average potential energy} = \rho \times A \times R^2 \times g/2T$$

where:

ρ = the density of water (1,000 kg/m^3)
A = the area of the reservoir
R = the mean tidal range
g = acceleration due to gravity (9.8 m s^{-2})
T = time

(Twidell and Weir 1986; Elliott 1996)

This equation, though, makes assumptions of constant basin area and total drainage of the basin and does not take other losses into account, and thus produces a greater value than would be achieved in practice.

Charlier states that only about 30 per cent of the potential energy can in fact be retrieved (Charlier 2003: 191), although he does not make it clear whether the 'energy losses' are from the low efficiency of the turbines at low head, the need for at least 2–3 m head height, or friction losses.

As Baker states, a greater total energy capture could be made through a larger number of turbines and generators able to pass a greater proportion of the flow over a shorter period, but this would prove uneconomic (Baker 1988: 20): 'Because the turbines would be operating for a very short time each tidal cycle, the energy produced per unit installed capacity would be small and the cost of the energy very high.' Additionally, the rapid release of large quantities of water 'would reduce the effective head across the turbines because there would be a slope in the water surface down towards the barrage on the basin side'. Baker states that there would also be various environmental problems resulting from the rapid release of water, presumably scour and turbulent 'white water' conditions.

In practice there is an optimum number of turbine generators for each site which represents a compromise between these factors and the need to utilise the turbines in an efficient and economic manner, i.e. retaining sufficient head height.

The output of the most recent 8.6 GW capacity Severn Tidal Barrage scheme is cited as 17 TWh p.a., and if only about 30 per cent of the potential energy is available as Charlier states, would imply a theoretical potential energy total of about 50 TWh. Turbine and generator losses, incomplete drainage of the basin and bed friction are assumed to account for the remainder of the potential energy.

The Severn Barrage would have 216 × 9 m diameter turbines rated at 40 MW each. The swept area of each turbine would be 63.617 m^2, hence

$$40{,}000 \text{ kW}/63.617 = 628 \text{ kW m}^{-2}$$

is the maximum power flux density at the turbine. The average power flux density would be 141.7 kW m^{-2} at the turbine. In practice, the power flux density would be greater since bulb turbines would be used, so reducing the effective swept area, and turbine and generator losses need to be factored in.

This value 628 kW m^{-2} would then represent the maximum power flux density after entrainment of the flow into the turbine tubes. The average power flux density would be lower, 141.7 kW m^{-2} at the turbine, since head heights vary between 4.4 m and 6.6 m under the ebb-only generation maximum energy output operation option (the most likely operating regime). Over the quarterly tidal cycle head heights vary between 11 m in the spring tide range and 4 m on the neap tide, i.e. by over half.

These power flux densities can be compared with those occurring without a barrage. Currents in the Severn Estuary at the site of the proposed barrage at Barry are cited as up to about 5.5 knots (nautical miles), or 2.8 m s^{-1} (ETSU 1993: 44), although elsewhere in the estuary up to about 8 knots or 4.116 m s^{-1} are said to occur. Assuming that the whole volume of the basin achieves this velocity, the kinetic energy per tide can then be calculated as:

$$KE = 0.5 \times 2.084 \times 10^9 \text{ m}^3 \times 4.116^2$$

where:

volume is 2.084×10^9 m^3 as cited by Elliott (1996)

$$KE = 1.7653 \times 10^{10} \text{ J}$$

which could roughly accord with the values given above, given that the tidal current velocity includes friction losses. As tidal energy is both kinetic and potential, only the kinetic component has been accounted for here, and the value would in fact be greater.

Per square metre, at a velocity of 2.8 m s^{-1}, the maximum kinetic power flux density before the barrage is constructed is:

$$0.5 \times 1{,}000 \times 2.8^3 = 10.98 \text{ kW m}^{-2}$$

or at 4.116m s-1 $\quad 0.5 \times 1{,}000 \times 4.116^3 = 34.87$ kW m^{-2}

On this basis the maximum power flux density is increased by entrainment by the barrage, from 10.98 kW m^{-2} to ~628 kW m^{-2}, or by ~57 times or by ~18 times for the higher current of 4.116 m s^{-1}. According to the hypothesis, this increase in maximum power flux density could result in impacts.

For currents of 8 m s^{-1} which occur elsewhere around the British Isles, e.g. Pentland Firth, and to almost this velocity at Alderney Race (in fact up to 10 m s^{-1} in some locations at Pentland Firth) (ETSU 1993), the power flux density would be:

$$0.5 \times 1{,}000 \times 8^3 = 256 \text{ kW m}^{-2}$$

The maximum power flux densities would be found at Pentland Skerries, where currents reach 16 knots or 8.2 m s^{-1} in the eastward direction (ETSU 1993: 30), which would be 275.68 kW m^{-2} power flux density, though this is exceptional.

The maximum power flux densities before and after construction of the Severn Barrage can be contrasted: 10.98 kW m^{-2}, which is relatively low, or up to about 34.87 kW m^{-2} and then after construction about 628 kW m^{-2} (in the turbine housing itself).

The proportion of the energy extracted has been stated by Charlier as ~30 per cent. Interception of the flow at the barrage would be nearly 100 per cent.

The number of conversions can be described as comprising:

- The conversion of the kinetic and potential energy of the 'tidal wave' behind the barrage into mainly potential energy, for the three or so hours of still water period at the height of the flood tide, so as to achieve the head height necessary for generation.
- The next conversion is from potential to kinetic energy, where the mainly still water is accelerated by gravity and its velocity increased, and also by the design of the turbine runner tubes up to a maximum of about 10 m s^{-1} (Baker 1988: 171) then being decelerated smoothly after the turbines to about 2 m s^{-1}. (Note that on this basis the maximum power flux density is 500 kW m^{-2} at the turbines.) There is a pressure drop to below atmospheric pressure at the turbines where the energy extraction occurs, but this recovers afterwards. Baker states that conversion back to potential energy occurs.
- Energy conversion of the flowing water from kinetic to the rotating turbine blades mechanical form occurs, and thence secondary conversion from mechanical to electrical in the generator.

The number of conversions of the flow can be said to amount to three; kinetic and potential to solely potential; potential to kinetic; and then back to potential. The energy extracted itself undergoes two further conversions, from kinetic to mechanical and mechanical to electrical.

At each stage there could be discontinuities in the functions of the energy flow and efficiency losses.

Regarding the efficiency losses: Baker states that the barrage, being highly porous with multiple sluices, would impede the flood flow only slightly. HEP turbines and generators are generally highly efficient at around 90 per cent, though for lower head heights this drops. For the ranges in question, i.e. around 3–7 m, lower efficiencies could be assumed, e.g. perhaps 60–70 per cent. For the overall scheme, it could be argued that the 30 per cent extractable energy cited by Charlier represents the efficiency, though this has been identified here as the *proportion* of the energy extracted.

7.3 Tidal lagoons

This concept is similar to that of tidal barrages, except that instead of employing a barrage to enclose an estuary in order to create a reservoir, the reservoir is created by embankments enclosing an area of shallow water, either set off at some distance from the coast and making no contact with it, or using the coastline as one boundary. The water of the high tide could either be retained at a level above that of the surrounding sea once the tide falls, and the potential difference in height used to turn a turbine and generate electricity in the normal ebb generation manner; alternatively, two-way generation could be employed. Inside the bounded area, the reservoir could either be a single basin, or be split into several pounds by further embankments or bunds. Each pound could then be kept at a different level so that continuous generation could be achieved. As yet no tidal lagoons have been built, although there are proposals for lagoons off the coast of South Wales and in China (Tidal Electric 2008).

There are various advantages and disadvantages to this approach. First, many of the impacts of tidal barrages can be avoided, for example the loss of inter-tidal mudflats through inundation. The tidal lagoon could be constructed in an area with continuous water cover, which it would still retain. Another advantage is that of either continuous generation for divided reservoir schemes or more continuous generation for ebb-and-flow generation. This is in contrast to the great bursts of power that an ebb generation tidal barrage produces, sometimes in the middle of the night when demand is at its lowest and thus the electricity is worth least. The barrier effect of the barrage on the estuary would be avoided and thus the impact to migratory fish. If the tidal lagoon does not make a landfall, there would be no effects on land drainage, or the water table. As the lagoon would not enclose an estuary, which is usually a precious environment due to its mudflats, shallow waters and part-fresh, part-saline water, as well as the high energy levels of tidal currents providing nutrient transport and mixing of waters, it could avoid interference with these processes and thus the associated impacts.

However, it cannot be assumed that tidal lagoons would be entirely impact-free. There are some disadvantages to the tidal lagoon concept. The ratio of the embankment or bund to the enclosed water volume could be higher (depending on size) than for a barrage, since the entire perimeter of the reservoir would be bounded, as well as the subdivision into different pounds. As a result, construction costs and impacts could be greater than tidal barrages per installed unit of power. However, the larger the lagoon is in scale the more water volume relative to length of perimeter embankment, since the area scales as a square. Very large lagoons may experience more leakage through sea floor and embankments. Undersea cables would be required to connect to the onshore grid, and these can be expensive, in the region of £1 million per kilometre (MCT 2008). Tidal lagoons would probably not be sited in the areas of highest tidal range, as these are found higher up in the estuary as a result of resonance effects from the shoreline shape. An example of this is that of the Severn Estuary, where the highest tides are to be found at the English Stones, near to the second Severn crossing (Baker 1988). Tidal range is an important factor in the economics of tidal energy and a reduced range would reduce energy generation, so increasing costs.

These are mainly economic issues associated with tidal lagoons; there are also environmental issues, which are described below.

7.3.1 Potential environmental effects of tidal lagoons

Impacts from construction could be potentially significant since the ratio of reservoir volume to embankment length would be increased. Embankments would consist of clay and sand hardened with a rock and then concrete coating (Tidal Electric 2008), although newer construction techniques envisage the use of geo-textile bags containing sand sucked from the sea bed. Increased turbidity from stirring up sediments might result from the depositing of thousands of tons to form embankments. A tidal lagoon would affect coastal processes as it might still obstruct the tidal flows of water into and out of nearby estuaries, although to a lesser degree than a tidal barrage, depending on its position and size relative to the estuary. Some changes to currents and possibly reduction of currents and reduced tidal range upstream in the estuary might be expected. The lagoon itself may have the effect of increasing some currents locally, as an obstruction in the estuary, as water flowed around the embankments of the bounded area. There would be a certain change to the sea and sea bed in the area of the reservoir, which might involve some sedimentation if the currents are lower than those of the pre-existing ones, and scouring and removal of sediments if currents increase. Water quality – e.g. salinity, oxygen content and turbidity – could be affected.

To a certain extent there would likely still be some wildlife effects for fish and marine animals and birds, as the reservoir area will be converted from open estuary to lacustrine conditions. Some of these may indeed be beneficial in terms of reduced wave and wind conditions, providing shelter compared

Table 7.2 Summary of potential tidal lagoon impacts

- Reduced tidal range – 50 per cent internally
- Tidal changes upstream – small
- Changed water currents
- Reduced/raised water speed – small?
- Potential siltation in bunded reservoir area
- Scouring possible
- Lacustrine environment created in reservoir area
- Changes to benthic environment
- Water quality effects
- Changed turbidity
- Fish pressure change damage – small?
- Wildlife changes – small?
- Construction impacts
- Loss of inshore fishing grounds?
- Creation of changed inshore fishing/aquaculture area
- Obstruction to shipping lanes/fisheries

to pre-existing conditions. However, fish may still be susceptible to some extent to damage by turbine blades or pressure changes when the reservoir pounds are discharging water for generation. Fewer fish are likely to be involved compared to estuarial barrages, as migratory paths would not be blocked and fish will find their way into the reservoirs accidentally, rather than intentionally. Tidal lagoons would also constitute obstructions to shipping lanes, fishing and recreational boats, and would require careful siting to minimise this. Lifetimes for tidal lagoons would be comparable with tidal barrages, e.g. up to 120 years (Tidal Lagoon Swansea Bay 2012) and the duration of any impacts might exacerbate cumulative effects. However, tidal lagoons could potentially result in lower impacts than barrages. Some potential tidal lagoon impacts are shown in Table 7.2.

The parameters of the impact model proposed in this book, that is the power flux density changes of proposed tidal lagoon schemes and their environs, could be considered, though this is not done here. The proportion of the energy in the flow extracted could be assessed, together with the number of conversions, the conversion efficiency and changes to the existing flow patterns, as has been done above for tidal barrages. Whereas impacts from one tidal lagoon scheme may be low, cumulative effects from a series of such developments along coastlines and estuaries might become more significant.

7.4 Tidal/marine current turbines

Marine current turbines are another method of harnessing tidal energy or other marine currents, but without a barrage. These turbines are somewhat similar to underwater wind turbines, and extract energy from the kinetic energy of the flow of the current rather than the potential energy of the tidal range stored behind a barrage, thus differing considerably from those in a tidal barrage. Generally, larger numbers of turbines would be required, dispersed through the flow channel. However, apart from the numbers required, there is much less civil engineering involved in installation compared to tidal barrages. Tidal current or marine current turbines are still a novel technology with few working examples, and these are mainly prototypes at the time of writing (MCT 2008). Nevertheless, there is worldwide interest with over 60 projects underway (Charlier 2003). Now, in the twenty-first century, the concept has been proven using propeller-type designs, both by the MCT turbines off Lynmouth and Strangford Narrows in Northern Ireland (MCT 2008), and those at Hammerfest in Norway (Hammerfest Strom 2008). Other designs have also emerged, such as the Open Centre Turbine developed by Open Hydro and oscillating hydrofoils systems (Stingray, Pulse Tidal) (NATTA 2006). In what follows, however, the discussion focuses just on the more common 'propeller' designs. While there are designs that employ ducts to concentrate the flows at the turbine, these are outside the scope of this study (Lunar Energy 2009).

It is envisaged that marine current turbines would be sited in 'farms' akin to windfarms, but immersed. In common with wind turbines, they would need to be dispersed to avoid each other's turbulent wake. Although it has been proposed to site such turbines in a 'tidal fence', i.e. intercepting most of the flow (Blue Energy 2008), it is not yet known whether this would prove feasible or indeed desirable. However, the proposed Severn tidal fence would only partially intercept the estuary (Parsons Brinkerhoff and Black & Veatch 2008).

At present, marine current turbine designs are generally limited to usually 20–40 m water depth and hub heights 10–20 m above the sea bed, since near the sea bed boundary-layer friction reduces the current significantly. Total immersion is required, taking into account tidal range and wave height, together with possibly a margin for shipping, imposing an upper limit (Bryden *et al.* 1998). A 20 m diameter is considered an upper limit due to cavitation effects for rotary turbine blades.

There will be an economic limit to the distance that marine current turbines can be sited away from the shore due to the cost of underwater power cables, which has been cited in the region of £160,000+ (at 1998 figures) per kilometre (Bryden *et al.* 1998), though up to £1 million per kilometre elsewhere.

Since the marine current turbines are taking energy from an 'extended fluid flow', the Betz limit of ~0.59 applies. This is the theoretical maximum energy that can be extracted from the flow before extra resistance becomes counterproductive by slowing the flow the turbine 'sees' to too great an extent (Twidell and Weir 1986: 216). Furthermore, the turbine power coefficient will apply. Since the flow velocity is relatively low, the turbine efficiency will tend to be less than that of high-head HEP turbines

Table 7.3 Potential environmental impacts of marine current turbines

- Removal of kinetic energy from coastal currents
- Tidal flow interaction
- Scouring effects on benthos
- Changes to benthic organisms' habitats
- Converter device installation disruption
- Fish strikes: low rpm (10–15 rpm)
- Inshore fishery effects
- Possible spawning ground disruption
- Visual impact
- Acoustic emissions
- Possible electromagnetic field (EMF) generation
- Coastline bird nesting site disruption
- Possible effects on tourism
- Highway access implications

(90 per cent plus). Therefore the proportion of energy that marine current technology can abstract is limited by the turbine power coefficient, including the Betz limit of ~59 per cent, and the spacing of turbines, preventing interception of the whole flow. The factors above limit the overall power flux density for energy extraction from the overall broad flow. Marine current turbines have low rotational speeds of 10–15 rpm (cited for the MCT converter), so fish strikes are less likely (Bryden *et al.* 1998; MCT 2008).

Overall environmental impacts could be assumed to be relatively low since apart from the removal of kinetic energy from coastal currents, there are no or few other fundamental changes incurred in the flow. Dacre (2002) has identified potential environmental impacts, as have MCT (2008) and Fraenkel (2007), but these will depend on the packing density of the siting. This could be expected to increase the impacts. The potential impacts are shown in Table 7.3.

7.4.1 Power flux density of tidal current schemes

The power density flux of a marine current turbine is considerably less than that of tidal barrages. For example, the maximum power flux per square metre of the marine current turbine off the coast of Lynmouth in Devon is ~9.8 kW m^{-2}, while the mean power flux density is ~3.7 kW m^{-2}. This has been calculated from a water speed of 2.7 m s^{-1} and a capacity factor of 0.375 (MCT 2008).

The maximum power flux density of the Severn Tidal Barrage is thus by this reckoning about 64 times that of the MCT tidal current scheme, although the difference in the two sites' currents and hence fluxes (the current velocity at the marine current site is one-third to one-half of that of the tidal barrage site) accounts for some of this difference.

Figure 7.1 Seagen Tidal Stream Marine Turbine at Lough Strangford N. Ireland in strong current. Image courtesy of Marine Current Turbines, a Siemens business. Copyright MCT ©.

Figure 7.2 Seagen Tidal Stream turbine close-up in strong current. Image courtesy of Marine Current Turbines, a Siemens business. Copyright MCT ©.

Figure 7.3 Seagen Tidal Stream turbine raised out of the water. Image courtesy of Marine Current Turbines, a Siemens business. Copyright MCT ©.

Bryden *et al.* have done work analysing the effect of tidal current energy extraction on channel flow as part of overall resource estimation and on wider effects, e.g. sediment transport (Couch *et al.* 2007; WEC 2007).

In extracting energy from flowing water in a channel, the kinetic energy of water is captured. The kinetic energy of flowing water can be expressed by the equation:

$$P(\text{in W}) = 1/2\rho A V^3$$

where:
ρ = density of sea water (kg m^{-3})
A = cross-sectional area of the flow (m^2)
V = velocity of flow (m s^{-1})

(Boyle 1996)

This equation is the same as that for wind energy. However, the density of seawater is about 833 times greater than that of air at sea level, while the velocity of marine currents will be in the range of 2–5 m s^{-1} rather than the 5–25 m s^{-1} of wind. Additionally, there are considerable differences between the broad and very high (or deep) flows of air in the atmosphere and the relatively shallow and limited-width channels where marine currents occur.

The common factors of laminar flow boundary friction from surface roughness and turbulent flow apply to both wind energy and marine current energy, but differ considerably in practice.

Marine current turbines operate by effectively slowing down the water flow, by a counter-resistance and thus the extraction of energy, i.e. the velocity is reduced (at least in the vicinity of the turbine). Tidal flows are driven by small variations in water height, or depth, and hence the water can be thought of as flowing down a (shallow) slope. A further effect of this slowing down of the current is to raise the water height, or depth, upstream of the turbine device with a corresponding resultant drop in the depth downstream (Couch *et al.* 2007).

Couch *et al.* have proposed a parameter for assessing the sensitivity of a channel to energy extraction, which employs the proportion, described as 'ratio', of energy extracted from a flow, combined with elements of Manning's equation:

$$B = \left[\frac{f}{1 + \frac{2gLn^2}{R^{4/3}}} \right]$$

Where:

B = sensitivity parameter
f = the ratio of energy extracted to the energy in the kinetic flow (i.e. proportion of energy extracted)
L = the channel length (m)
g = the acceleration due to gravity (m s^{-2})
n = Manning's roughness coefficient
R = the hydraulic radius (m)

(Couch *et al.* 2007)

This equation employs the concept of the proportion of energy extracted in the kinetic flow, termed here as the ratio, a parameter proposed by this author (Clarke 1993).

7.4.2 Tidal current resource estimates

Both Salter (2005) and MacKay (2007) suggested that the resource estimates for tidal currents have been considerably underestimated. Salter proposed that bed surface roughness dissipates a large proportion of the energy of the tidal flows of the Pentland Firth, between the north of Scotland and the Orkney Islands, and that marine current turbines would extract energy that would have previously been dissipated in turbulence. The horizontal water flow velocity would then remain relatively unchanged, with lower effective energy loss, since the flow would now be much less turbulent and of a laminar nature, it is argued.

This may well be the case; however, Salter appears to be considering the availability of energy that can be captured from the tidal flow and not the environmental functions of that energy and its implications. If as much as 100 GW peak is dissipated in the Pentland Firth area by turbulence and surface roughness, as Salter suggests, the question of what that energy is doing and whether it is environmentally significant needs to be posed.

To be environmentally significant, such functions as scouring, lifting and entraining of sediment by the energy flow could be sought. Turbulent flows, involving eddies of high velocity and pressures, can be very effective agents of scouring and erosional processes (Graf 1971). Turbulence, consisting of vortices, can have locally high power density fluxes.

If this proves to be the case, then abstracting energy from the flow and reducing the turbulence there might have significance for the lifting and transport of sediment and deposition elsewhere. Of course, it may transpire that the sea bed of the Pentland Firth is composed of scour-resistant beds, and that there are no significant volumes of sediments being entrained and transported.

However, it cannot be assumed that because the energy is dissipated in turbulent flows that it is then 'incidental' and has no function in the local environment. On the other hand, where such flows perform functions, extraction of a proportion of the energy available, without significant impacts, may be possible. The determination of the critical proportion that can be extracted would be required, through modelling exercises and the use of sensitivity analyses, such as that of Couch *et al.* (2007).

MacKay (2007) also thought the overall resource might have been significantly underestimated, but for a different reason. MacKay suggested underestimation of the resource (MacKay 2007) due to incorrect estimation of the power available, from considering only the kinetic energy involved in the flow, rather than assessing the resource in terms of a 'tidal' wave. Wave energy has both kinetic and potential energy components (Duckers 1996). He points out that Black & Veatch (Parsons Brinkerhoff and Black & Veatch 2008), who carried out the resource estimation for the Carbon Trust, have employed only the kinetic component, i.e.

$$P(\text{in W}) = 1/2\rho A V^3$$

as cited above. This formula was used to obtain a figure of 12 TWh yr^{-1} as a total UK tidal current estimate. MacKay, however, states that 'the power in tidal waves is not equal to the kinetic energy flux across a plane'. He goes on to demonstrate mathematically that this is not so by comparing the power estimates from the kinetic energy equation with those from using the wave power equation. He states that the potential energy of a wave (per wavelength and per unit width of wave front) is:

$$\text{PE} = 1/2\rho g h^2 \lambda$$

Where:
ρ = density of seawater
g = gravity
h = wave height
λ = wave length

This equation is somewhat similar to that given by Duckers (Duckers 1996: 323) for the power in watts per metre width of wave front of a pure sinusoidal wave:

$$P = \frac{\rho g^2 H^2 T}{32\pi}$$

Where:
P = power
ρ = density of seawater
g = gravity
H = wave height
T = wave period

7.4.3 Applying the hypothesis to marine current turbines

The maximum power flux density in terms of the kinetic energy available in the current at the converter will be in the range of 2.7 kW m^{-2} to 72.35 kW m^{-2}, based on currents of between 1.75 m s^{-1} and 5.25 m s^{-1} (MCT 2008; ETSU 1993: 5–7). However, in practice the very fast currents occur infrequently only in certain locations. Typical ranges might be more like that of the Lynmouth MCT turbine maximum of ~9.8 kW m^{-2} and an average of ~3.7 kW m^{-2}.

The proportion of the energy in the flow that can be extracted will, as stated, be limited by the power coefficient Betz upper limit of 0.59, and proportion intercepted by the spacing of turbines, though given the viscosity of water, the effect of the turbines will be to slow a water column cylinder of greater diameter than the turbine itself. The proportion of the flow intercepted would perhaps not be much more than about 25–30 per cent, if turbines were stacked for maximum effect, unless a 'tidal fence' arrangement or successive arrays of turbines were sited downstream and upstream.

The device will probably not capture more than about 40–50 per cent of the energy in the flow it 'sees' overall, due to turbine power coefficient, cited as >0.45 (Fraenkel 2007), and gear train and generator losses.

The number of energy conversions and changes to the flow is limited to a reduction in velocity, plus some angular change; though a corkscrew wake pattern will result directly downstream of the turbine, which will dissipate itself in a widening of the water column affected. Since most of the kinetic (and potential) energy is not converted into either one or the other form, as with tidal barrages, fewer impacts might be expected. Energy conversions are from the kinetic energy of the water, to the rotational mechanical energy of the blades and shaft and thence via the gearbox to electrical conversion in the generator. The smaller number of conversions could imply fewer impacts according to the hypothesis.

7.4.4 Comparison of tidal barrage (range) and tidal current (flow) technology

It would appear that a tidal barrage has the potential to capture a greater proportion of the energy available than do marine current turbines; this is achieved (for ebb flow generation) by entrainment and capture of the water and its potential energy at high tide point. This energy is then converted by opening sluices into kinetic energy under the influence of gravity to turn the turbines and generator. By comparison, a marine current turbine converts some of the kinetic energy of the water directly into rotary (or oscillatory) mechanical motion of the blades (or hydrofoils) and so turns a generator via a gear train or hydraulic drive.

The environmental impacts of tidal barrages have been well documented and on occasion are cited as one of the reasons for not proceeding with construction (Clarke 1995). The case of tidal barrages illustrates how a number of energy conversions, i.e. from kinetic to potential energy, then back to kinetic, can have environmental consequences. The holding of the water volume at the high-tide point for a period in order to achieve a useful head height, i.e. potential difference, also arrests the current and with the drop in current speed, transported sediments will be deposited (STPG 1986; Shaw 1980). In effect the oscillatory flow of the water is being accentuated by being periodically stored then released in sharper bursts.

A result of retaining the water of the flow is that the tidal range inside the barrage is reduced by about 50 per cent or less (STPG 1986). The change in the estuary to a semi-lacustrine environment is caused by this water retention. This can be considered as one of the major impacts of tidal barrage schemes, and represents the greatest change from inter-tidal mudflats to a continually immersed 'lower energy' environment, depending on the site bathymetry. Impacts such as fish or eel strikes could be connected to pressure change across the turbine. That is, the higher power flux density and pressure, and the greater the changes to these conditions, the greater the likelihood of impacts.

By contrast, a tidal current turbine is likely to have much less impact; the blades rotate more slowly at 10–15 rpm (MCT 2008), so there is less risk to fish, power flux density is much lower and it is not necessary for estuaries to be blocked, so there is likely to be much less impact on tidal range or flooding.

However, there could be some minor impacts in terms of, for example: energy extraction, though lower than for barrages; visual intrusion if they have an above-water aspect; and constraints on shipping movements and fishing. Installation involves potential impacts to marine life, and in operation there may also be some impacts on sea mammals, e.g. due to low-frequency noise (MCT 2008).

This suggests that the lower power flux densities of marine current turbines, and the reduced changes in power flux density that are involved, could lead to reduced impacts.

7.5 Wave energy

Wave energy conversion has a little more developmental experience than marine current turbines, though it is barely beyond the prototype stage. The first commercial offshore wave energy converters were commissioned in Portugal in 2008 (Pelamis 2009), though decommissioned soon after due to problems, and there are more than 15 years of experience with prototype onshore wave devices. The variety of device designs indicates that the technology is still relatively immature. Although most projects have their antecedents in the 1970s, progress has been intermittent due to inconsistent government support.

7.5.1 The wave energy resource

The UK's potential wave energy resource is substantial, the UK being one of seven countries in Europe with significant wave resources. Europe's wave energy resource has been estimated at 50 GW total power in waves kW m^{-1} (Mollison 1989), while the UK has an estimated 12 GW, or a 7–10 GW (technical potential) (ETSU 1993), which would be equivalent to ~20–25 per cent of electricity supply. In practice, constraints would reduce the amount available. The total world resource has been estimated at 1–10 TW. At 2 TW this would be equivalent to approximately twice world electricity generation (DTI 2004; WEC 2004).

For the UK, offshore resources of 30–90 kW per metre of crest width constitute the major part of the total resource, with the maximum output 10 km offshore (DTI 2004). The size of waves depends on wind speed, duration of wind, and length of fetch – the distance over which the wind has blown (Duckers 1996: 320). As a consequence, the European resource occurs mainly on the western Atlantic coast. Waves near the shore dissipate much of their energy in friction with the sea bed, which eventually causes waves to break.

The challenge of harnessing wave energy lies partly in the fact that waves have very variable patterns of length, height and period, as well as direction. The structural forces can be up to 100 times greater than the average in storms (Twidell and Weir 1986: 312), and the device must survive this hostile environment. The irregular slow motion of waves, e.g. of 0.1 Hz frequency, needs to be transformed into 500 times the frequency for generators, requiring elaborate gear train and accumulators or conversion into other forms to produce an even energy output.

7.5.2 Environmental impacts of wave energy

The characteristics of wave energy and wave energy conversion contribute to the environmental effects. Wave action on the shorelines can be important agents of functions in the natural environment, such as erosion and deposition, as well as oxygenation of water and contribution to currents. Beach shapes are the consequence of wave action, and beaches change in shape from summer to winter due to greater wave action in the winter period (ETSU 1979).

Although there are few or no environmental impacts recorded as yet from the handful of wave converters deployed, since the impacts would be at almost imperceptible levels, some effects can be anticipated as deployment increases.

A CUC/UEC study of the impact of wave energy converters from the South West Wave Hub on surf heights in Cornwall found only a small reduction in energy levels at the shore. A maximum change of 2.3 per cent, or 4 cm, for an energy transmission rate of 90 per cent, i.e. the proportion of energy extracted at the wave energy converter was 10 per cent (Millar *et al.* 2007). More realistically average changes were in the region of 1 cm in significant wave height. The study considered energy transmission ranges of 0 per cent, 70 per cent and 90 per cent for a reference array of wave converters of ~3,900 m, with seven 300 m long units spaced out at 300 m intervals (Millar *et al.* 2007). The Wave Hub is an experimental 'plug in' facility so that prototype wave machines can be tested at sea in offshore wave conditions. A commercial wave farm would perhaps be similar or larger in scale and there would be numbers of them, so that more of the incident wave energy would be extracted and greater effects on shore could be expected.

Wave-energy converters can be expected to reduce waves in their lee, even if only slightly, as they extract energy from the waves. Onshore devices would involve changes and modifications to the shoreline, and there would be direct impacts from construction at those locations. Offshore devices do not suffer from this characteristic, though tethering and anchoring on the sea bed are needed as well as cabling connections to the shore. Some near-shore devices would be sea-bed mounted, however.

A certain amount of work has been done on the potential environmental impacts of wave energy, which has been reviewed by this author (Clarke 1995). A summary of the main anticipated impacts is given in Table 7.4.

7.5.3 Applying the hypothesis to wave power

Power flux density for wave energy

In terms of power flux density, wave energy has been cited as having a 20–40 kW m^{-1} mean power density in the UK context (Clarke 1993).

Table 7.4 Potential environmental impacts of wave energy

Onshore devices

- some land use
- some visual
- noise
- cable landfall/grid connection

Near-shore/offshore devices

- visual
- safety
- shipping obstruction
- wildlife
- anti-fouling coating
- cable landfall/grid connection
- geomorphic effects from energy extraction: beach and shore changes possible
- changed longshore currents
- possible underwater noise effects on wildlife

Source: Clarke (1995)

Translating this value of kilowatts per metre to kilowatts per square metre requires sea depth and wavelengths data. In Clarke (1993), wave depths of 10 m were assumed and the mean value of 20–40 kW m^{-1} was divided by the wave depth to obtain the values cited of ~3 kW m^{-2} and a maximum of ~50 kW m^{-2}. Since 95 per cent of the energy is in the top layer to a depth of one-quarter (Duckers 1996: 326) of the wavelength, the power flux density can be cited as roughly four times this amount, i.e. 30 kW m^{-1}/2.5 m for a 10 m wave depth = 12 kW m^{-2}.

Waves have an orbiting water 'particle' motion and the orbits decrease in size exponentially with depth (Duckers 1996: 326). A difficulty in assessing the power flux density per square metre arises because wave energy is in part kinetic energy and in part potential energy (Duckers 1996).

Power in waves

The equation for the power in watts per metre width of deep water wave for an idealised sinusoidal wave is given by Duckers (1996):

$$P = \frac{\rho g^2 H^2 T}{32\pi}$$

Where:

P = power (in watts)
ρ = density of seawater (1,025 kg m^{-3})
g = gravity (9.81 m s^{-1})
H = wave height (m)
T = wave period (s)

As a general approximation:

$$P \text{ (in kW)} = H^2T \text{ (in seconds)}$$

(Duckers 1996)

since the value of ρ is ~1,000, and g^2 is ~96.2 and 32π is ~100.5. Deep water is defined as greater than half of the wavelength.

Hence the power in the waves increases as a square of the wave height and linearly with increases in the wave period. Long-wavelength-type waves or swell, produced over long reaches, contain the most power.

For real seas, which have varying waves, the significant wave height or H_s is used, i.e. the average of the highest one-third of the waves (Duckers 1996).

The power can be estimated as:

$$P = 0.55\,(H_s)^2\,T_z$$

Where:
P = kW m^{-1}
H_s = significant wave height in metres
T_z = wave period in seconds

(ETSU and CCE 1992)

Types of wave energy converters

Applying the hypothesis to different types of wave energy converters, the power flux density of the main categories can be compared. Duckers groups wave energy converters into three types: *terminators* which face the wave front; *attenuators* which are perpendicular to the wave front; and *point absorbers* which have small dimensions and absorb energy from surrounding waves (Duckers 1996: 329). Further categories are: onshore, near-shore and offshore. Some designs of wave energy converters concentrate waves, amplifying them in the process, such as the tapered channel converter or Tapchan. This latter type stores the potential energy of the wave in a reservoir some two or so metres above the mean sea level, and uses a low-head turbine to generate constant power.

Terminator designs seek to extract the maximum energy from the wave by absorbing it head-on and leaving flat water behind. An example of this approach is Salter's 'Duck' design, which could achieve high-efficiency energy conversion of over 90 per cent in suitable conditions. Lines of Salter's Ducks were proposed for large-scale wave energy developments off the north-west coast of the UK.

Power flux density reduction onshore

Power flux density of 67 kW m^{-1} of wave front is found offshore from the Outer Hebrides in deep water (Mollison 1989). If it is assumed that

95 per cent of this energy is in the top 2.5 m of 10 m deep waves, this would equate to ~25 kW m^{-2} power flux density. By contrast, only a fraction of this energy per metre would reach the shoreline due to bottom friction. Duckers states that

> waves of 50 kW m^{-1} in deep water would have only 20 kW m^{-1} or less when they are closer to the shore or in shallow water, depending on the distance travelled in shallow water and the roughness of the sea bed.
>
> (Duckers 1996)

Much lower power flux densities occur at onshore wave energy converters than for offshore. Examples of onshore wave energy converters exist, such as the Limpet 500 kW Oscillating Water Column device on Islay. Here, the power flux density of the waves was assessed at only about 20 kW m^{-1}, but in practice proved to be less (Queens University Belfast 2002), and the device has proved capable of only about 150 kW or less of output (Whittaker *et al.* 2005). The energy of the original deep water waves has been absorbed by friction with the bed (Duckers 1996), resulting in shear and turbulent forces with locally high power flux density that can pick up sand and sediments and further abrade the sea bed. The erosion and sediment-mixing processes that occur through this energy dissipation constitute some of the functions of wave energy in nature.

It is suggested, then, that although onshore wave converters result in other impacts, such as changes to the shore, fewer impacts from removal of energy from the waves may result from this type of converter, since much of the energy has been dissipated already as the wave travels through shallow water.

Proportion of flow extracted

For offshore wave converters, how much of the wave's energy reaches the shoreline will depend on the proportion extracted, which will in turn depend on the efficiency of the device and the interception ratio. Salter's Ducks were conceived as long lines of converters attached to each other by a spine, offshore in deep water facing the waves. Assuming that high-efficiency energy extraction were to be achieved, such long lines with few gaps could conceivably change the wave environment at the shore considerably (ETSU 1979). However, gaps would be required for shipping, fishing and other access. Additionally, it is unlikely that as much as 90 per cent of the wave energy could be extracted, since wave length and height vary considerably (Duckers 1996: 324). Third, the highest power flux densities would occur during storms, with the result that the majority of the natural functions would in fact be performed then. The energy of the waves would be too great to be captured by the device since such devices would be matched to the most productive mean wave ranges per annum (i.e. balancing frequency with productivity) wave regimes rather than the infrequent highest.

In practice, wave energy technology has not yet matured to the point of developing offshore terminator devices; having started with onshore devices, it has moved to near-shore and more recently to offshore attenuator devices with, e.g. the Pelamis device. Survival during storms is a considerable challenge for offshore terminator devices.

Offshore Pelamis device: an example

The Pelamis wave energy converter consists of several 3.5 m diameter, 30 m long cylindrical sections linked by hinged hydraulic pump joints, which pump high-pressure oil to hydraulic motors to drive a generator (Pelamis 2009). The attenuator Pelamis units, 120 m long, lie parallel to the wave direction, i.e. perpendicular to the wave front and by design aim to capture a lower proportion of the energy available than does Salter's Duck.

The original Pelamis units were rated at a maximum output of 750 kW each. At their first commercial application site in Portugal, at Agucadoura, 8 km offshore, the 2.25 MW installation of three units, plus a further 20 MW expansion planned, was expected to supply power for 1,500 households (Pelamis 2009). At ~1 kW continuous demand from the households, this would imply an average output of 1 × 1,500 = 1.5 MW. The capacity factor would then be ~0.067, or about 6.7 per cent.

Calculation of proportion of wave energy extracted by Pelamis development

The wave power level is 46 kW m^{-1} offshore of the north-west of Portugal (CRES 2002).

- Assuming wave front = 1 km
- Total power in waves per km^2 = 46 MW
- At a packing density = 30 MW km^{-2}

At a capacity factor of 0.067, as assumed for the Agucadoura scheme, the ratio of wave energy converted into electricity per kilometre would equal 30 × 0.067 = ~2 MW/46 MW = 0.043, or about 4.3 per cent.

The efficiency of the Pelamis energy conversion chain, from hydraulic pumps to hydraulic generator and electrical generator, as well as cable connections and transformer, would need to be factored in.

Assuming an average of about 50–60 per cent efficiency (the actual efficiency of conversion would vary with wave height and period), and since lower efficiency results in greater interception of energy for a given output,

0.043/0.6 = 0.07

On this basis the proportion of average wave energy extracted at this location would be approximately one-fourteenth for the Pelamis device. More sophisticated models could produce more precise figures.

This would be a lower proportion of energy extracted than for terminator devices, and depending on the overall packing density for the shoreline, could be expected to result in fewer impacts.

Studies carried out by Wave Dragon, another type of offshore wave energy converter, have indicated that 'wave heights are expected to be reduced by 22%–32% 1 km behind a Wave Dragon farm', though this will depend on local conditions (Wave Dragon 2009). Presumably this reduction would decrease to some extent with further distance, depending on the proportion of wave front interception.

Efficiency of conversion

Salter's Ducks could achieve ~90 per cent efficiency in optimal conditions (Thorpe and Marrow 1990). The Tapchan scheme planned for Mauritius is cited by Twidell and Weir as achieving a theoretical 30 per cent efficiency from wave to electrical busbar (Twidell and Weir 1986: 334). The Oscillating Water Column (OWC) device efficiency will tend to be reduced by the conversion to compressed air, since this will produce some heat, air as a gas being compressible; furthermore, the turbine itself and duct will result in more losses.

Capture efficiency and power chain efficiency estimates were given in the Wave Energy Review 1990 (Thorpe and Marrow 1990) for several wave converter devices: the NEL Breakwater, NEL Floating Terminator, NEL Floating Attenuator, Vickers Terminator, Vickers Attenuator, Belfast Device, SEA Clam, the Lancaster Flexible Bag and the Edinburgh Duck and Bristol Cylinder. This is shown in Table 7.5.

Table 7.5 Capture and power chain efficiency and availability for several wave converter devices

Device	*Proportion captured*	*Power chain efficiency*	*Availability*
NEL Bottom standing OWC	Over 50%	55–63%	90%
S.E.A Clam	0.62 (derived)*	59–65%	70%
Edinburgh Duck	0.93 (derived)*	70–82%	92%

Source: Thorpe and Marrow (1990).

Note

* Based on (kWm^{-1} × device length)/(average annual generation/power chain efficiency/availability)

Number of energy conversions

The wave energy converters of terminator design, such as the Duck or the Clam, attempt to harness the energy in one conversion from the wave, resisting the force mechanically head-on without transforming the wave into another form or entraining it. The Tapchan devices concentrate the wave, channelling it up a flume of resonant dimensions; the flow overtops the reservoir, where it is converted to solely potential energy. Thereafter the water falls under the force of gravity to attain kinetic energy and turns a turbine. The number of energy conversions of the flow is three: from the kinetic and potential of the wave to potential only; then to kinetic energy; and then extraction of kinetic energy. The process might be compared to wave action in rock pools.

The OWC wave energy converter channels the wave into a tapering chamber or cylinder, which then compresses air (Duckers 1996). The air is pushed through a turbine generator. When the wave column falls, air is sucked back through the turbine. By using a dual-direction turbine with symmetrical blades (Wells turbine), continuous rotation in one direction can be achieved, without rectification of the air flow. Energy conversions comprise the change in kinetic and potential wave energy to compressed air, then to kinetic energy of the air passing through the turbine, and so rotational energy, i.e. three conversions. There will be some concentration of the wave into the narrow chamber. The wave itself undergoes only one conversion, the change from the kinetic and potential energy of the wave to a potential-only form, as the vertically rising and falling piston of the water column, although the water would flow out again, indicating conversion back to a largely kinetic form.

For the Pelamis device, the number of conversions of the wave is only one, transforming some of the kinetic and potential energy of the wave to mechanical motion of the device, which by resistance reduces the wave height and may affect the period. The effect would be to attenuate the wave rather than to produce calm water to the landward side.

7.6 Conclusion

The hypothesis has been extended to other water-based renewable sources and this suggests ways in which impacts could be linked to the power flux density of these sources, as well as the further factors employed here. As yet there are too few wave and marine current converters in existence for firm evidence of impact to have been gathered. A comparison between tidal barrage and marine current turbines demonstrates how the hypothesis could be applied. However, it is outside the scope of this book to verify this with tests of data linking impacts to power flux density. The paucity of tidal barrages, marine current and wave energy schemes worldwide, as yet, makes it difficult to collect actual data. Tidal barrages have much higher power flux density at the converter than marine current turbines, potentially up to about 500–628 kW m^{-2} (proposed Severn Barrage scheme), as compared to marine current turbines' average of ~10 kW m^{-2} , and this would appear likely to be reflected in the impacts likely to be experienced, which appear to be lower for marine current turbines.

The proportion of the energy flow intercepted is greater for tidal barrages, at ~100 per cent, compared to marine current turbines at probably a maximum of about 30 per cent or less, although the Tidal Fence schemes may intercept higher proportions.

The proportion of the energy abstracted has been cited at 30 per cent for tidal barrages; for marine current turbines it would probably be lower, depending on the scheme. The number of conversions of form of energy in the flow is lower for marine current turbines – only one compared to three for tidal barrages – and this could be expected to result in fewer impacts and disruptions of existing functions and conditions, such as sediment movement and tidal range.

The efficiency of energy conversion, at the converter would be higher for tidal barrages, as the low-head HEP-type turbines could be expected to be ~65–85 per cent efficient, compared to the marine current turbine efficiencies which would be limited by their overall coefficient of performance, C_p, up to the Betz limit, e.g. about 45 per cent, in common with other turbine types.

For wave energy converters, average power flux densities of ~12 kW m^{-2} for offshore wave energy converters on the west coasts of the UK and Ireland can be cited based on 20–40 kW m^{-1} average power levels per metre of wave front. At least 50 per cent of this would have been dissipated in bottom friction near-shore and up to 80 per cent by the shoreline. Since the dissipation of this energy represents the natural functions performed by waves listed above, higher power flux densities could be expected to give rise to more functions, and a diminution of them through extraction of energy could be likely to result in impacts. Such impacts as changes in wave height and beach shape, and related geomorphic and biological effects, as well as water mixing and oxygenation, may be anticipated. Currently studies envisage that less than 10 per cent of the energy will be extracted by offshore devices, and with present anticipated deployment densities this will incur only small changes at the shore, e.g. the south-west Wave Hub maximum of 2.3 per cent change at the shore. However, greater deployment rates would increase this. Wave Dragon converter studies indicted that a 22–32 per cent reduction in energy levels could be expected 1 km behind a Wave Dragon farm, and even sited further offshore significant recharging may not be possible as the effective fetch would be limited to several kilometres.

The interception ratio and the proportion of energy extraction appear likely to be relevant factors in overall impact. The proportion extracted relies partly on the efficiency of energy conversion by the device. Currently onshore and attenuator devices have estimated energy capture ratios of

~50–93 per cent, though in practice actual devices may have less. Terminator designs could be expected to have high interception ratios and thereby high energy extraction ratios.

The number of conversions could prove to be a significant factor in impacts. This varies from about three in the case of the Tapchan and OWC devices, to one in the case of Pelamis and other undeveloped designs such as Salter's Ducks.

Therefore the above suggests that the hypothesis might be usefully extended to other water-based renewable energy sources, if the environmental impacts (e.g. beach shape, sediment erosion and transport, as well as use of sea area) prove to be related to the parameters proposed – power flux density, proportion of interception and energy capture, efficiency of energy conversion, and number of energy conversions – in a manner analogous to HEP.

Bibliography

Baker, A.C. (1988) *Tidal Power*, Peter Peregrinus Ltd for IEE, Stevenage.

Bernstein, L.B. (1961) *Central Tidal-Power Stations in Contemporary Energy Production*, State Publishing House, Moscow.

Blue Energy (2008) 'Blue Energy', online resources: www.bluenergy.com, accessed 18 November 2008.

Boyle, G. (ed.) (1996) *Renewable Energy Power for a Sustainable Future*, Oxford University Press in conjunction with the Open University, Milton Keynes.

Bryden, I.G., Naik, S., Fraenkel, P. and Bullen, C.R. (1998) 'Matching Tidal Current Plants to Local Flow Conditions', *Energy*, Vol. 23 No. 9: 699–709.

Charlier, R.H. (2003) 'Sustainable Co-Generation from the Tides: A Review', *Renewable and Sustainable Energy Reviews*, Vol. 7: 187–213.

Clarke, A. (1993) 'Comparing the Impacts of Renewables: A Preliminary Analysis', TPG Occasional Paper, Technology Policy Group 23, Open University, Milton Keynes.

Clarke, A. (1995) 'Environmental Impacts of Renewable Energy: A Literature Review', Thesis for a Bachelor of Philosophy Degree, Technology Policy Group, Open University, Milton Keynes.

Couch, S.J., *et al.* (2007) 'Survey of Energy Resources', WEC Survey of Energy Resources 2007, online resources: www.worldenergy.org/documents/ser2007_final_online_version_1.pdf, accessed 6 August 2009).

CRES (2002) 'Wave Energy Utilisation in Europe: Current Status and Perspectives', online resources: www.cres.gr/kape/pdf/download/Wave%20Energy%20Brochure.pdf, accessed 10 June 2009.

Dacre, S. (2002) 'A Scoping Study to Establish a Research Programme on the Environmental Impacts of Tidal Stream Energy', Department of Trade and Industry, Contractor Robert Gordon University Aberdeen, Energy Technology Support Unit.

Department of Energy (1989) *The Severn Barrage Project: General Report*, HMSO, London.

DoE (2009) 'Report to Congress on the Potential Environmental Effects of Marine and Hydrokinetic Energy Technologies', US Department of Energy, Wind and

Hydropower Technologies Program, December, online resources: http://energy.gov/eere/water/downloads/report-congress-potential-environmental-effects-marine-and-hydrokinetic-energy, accessed 30 November 2015.

DTI (2004) 'Atlas of UK Marine Energy Resources', Department of Trade and Industry, Business Enterprise and Regulatory Reform Department, BERR, online resources: www.renewables-atlas.info, last accessed 6 August 2009.

Duckers, L. (1996), 'Wave Energy', in Boyle, G., ed., *Renewable Energy Power for a Sustainable Future*, Oxford University Press in conjunction with the Open University, Milton Keynes.

EC Birds Directive (1979) 'EC Birds Directive 79 /409/ EEC', European Union, online resources: http://ec.europa.eu/environment/nature/legislation/birdsdirective/index_en.htm, accessed 18 November 2008.

Elliott, D. (1996) 'Tidal Power', in Boyle, G., ed., *Renewable Energy Power for a Sustainable Future*, Oxford University Press in conjunction with the Open University, Milton Keynes.

ETSU (1979) *Wave Energy*, HMSO, London.

ETSU (1993) 'Tidal Stream Energy Review Appendices', Engineering & Power Development Consultants Ltd with Binnie & Partners Sir Robert McAlpine & Sons Ltd, and IT Power Ltd, ETSU on behalf of Department of Trade & Industry, Energy Technology Support Unit.

ETSU and CCE (1992) 'An Assessment of the State of the Art, Technical Perspectives and Potential Market for Wave Energy', ETSU-OPET for the DG17 European Commission.

Ford, N. (2006) 'Seoul Leads Tidal Breakthrough', *International Water Power & Dam Construction*, September, online resources: www.waterpowermagazine.com/story.asp?sc=2038863, accessed 8 June 2009.

Fraenkel, P. (2007) 'Marine Current Turbines: An Update', MCT Ltd online resources: www.all-energy.co.uk/UserFiles/File/2007PeterFraenkel.pdf, accessed 18 November 2008.

Gibrat, R. (1966) 'L'énergie des marées', *Rev. Franc. de l'Energie*, Vol. 183: 660–84.

Graf, W.H. (1971) *Hydraulics of Sediment Transport*, McGraw-Hill, New York.

Hammerfest Strom (2008) 'Norwegian Technology for Tidal Energy to Be Further Developed in Great Britain', Hammerfest Strom, online resources: www.hammerfeststrom.com/content/view/49/72/lang, accessed 19 November 2008.

Hubbert, M.K. (1969) *Resources and Man*, Freeman, San Francisco, CA.

Joint Nature Information Committee (2008) 'Information Sheet on Ramsar Wetlands (RIS) Severn Estuary', Joint Nature Information Committee, online resources: www.jncc.gov.uk/pdf/RIS/UK11081.pdf, accessed 6 August 2009.

Lunar Energy Ltd (2009) Homepage, online resources: www.lunarenergy.co.uk, accessed 8 June 2009.

MacKay, D.J.C. (2007) 'Under Estimation of the UK Tidal Resource', Cavendish Laboratory, University of Cambridge, 17 April 2007, online resources: www.inference.phy.cam.ac.uk/sustainable/book/tex/TideEstimate.pdf, accessed 6 August 2009.

MCT (2008), 'Marine Current Turbines', online resources: www.marineturbines.com (accessed 18 November 2008).

Millar, D.L., Smith, H.C.M. and Reeve, D.E. (2007) 'Modelling Analysis of the Sensitivity of Shoreline Change to a Wave Farm', *Ocean Engineering*, Vol. 34: 884–901.

Mollison, D. (1989) 'The European Wave Power Resource', UK ISES Conference on Wave Energy Devices, Lanchester Polytechnic, 28 November.

Mosonyi, E. (1963), 'Utilizable Power in Seas and Oceans', in *Water Power Development I*, State Publishing House, Moscow.

NATTA (2006) 'Wave Power and Tidal Current Turbines', Network for Alternative Technology and Technology Assessment, c/o Energy and Environment Research Unit, Open University, Milton Keynes.

Parsons Brinkerhoff and Black & Veatch (2008) 'Severn Tidal Power Feasibility Study', Parsons Brinkerhoff and Black & Veatch, online resources: www.severnestuary.net/sep/forum/tidalpoweroptions.ppt, accessed 18 November 2008.

Pelamis (2009) Wave Power Ltd Company Website. Online resources: www.pelamiswave.com, accessed 9 June 2009.

Queens University Belfast (2002) 'Islay Limpet Wave Power Plant Publishable Report', Queens University Belfast and EC Joule III, online resources: www.wavegen.co.uk/pdf/LIMPET%20publishable%20report.pdf, accessed 10 June 2009.

Rasmussen, J. N. (2015) 'Vindmøller slog rekord i 2014' Energinet dk, 6 January 2015. Online resources: http://energinet.dk/DA/El/Nyheder/Sider/Vindmoeller-slog-rekord-i-2014.aspx, accessed 5 September 2015.

Salter, S.H. (2005) 'Possible Under-Estimation of the UK Tidal Resource', Submission for Question 2, online resources: www.berr.gov.uk/files/file31313.pdf.

Shaw, T.L. (ed.) (1980) *An Environmental Appraisal of Tidal Power Stations: With Particular Reference to the Severn Barrage*, Pitman Publishing, London.

STPG (1986) 'Tidal Power from the Severn Interim Study Report', Severn Tidal Power Group, London.

Sustainable Development Commission (2007) 'Turning the Tide: Tidal Power in the UK', Sustainable Development Commission, online resources: www.sd-commission.org.uk/publications/downloads/Tidal_Power_in_the_UK_Oct07.pdf, accessed 18 November 2008.

Thorpe, T.W. and Marrow, J.E. (1990) 'Wave Energy Review Progress Report 1', Chief Scientists Group, Energy Technology Support Unit.

Tidal Electric Limited (2008) 'UK Projects', Tidal Electric Limited online resources: www.tidalelectric.com/Projects%20UK.htm, accessed 6 August 2009.

Tidal Lagoon Swansea Bay (2012), 'Proposed Tidal Lagoon Development in Swansea Bay, South Wales Environmental Impact Assessment Scoping Report', Tidal Lagoon Swansea Bay, online resources: www.tidallagoonswanseabay.com/document-library/environmental-scoping/95, accessed 30 November 2015.

Twidell, J. and Weir, A. (1986) *Renewable Energy Resources*, E&FN Spon, London.

Wave Dragon (2009) 'Wave Dragon: Environment', online resources: www.wavedragon.net/index.php?option=com_content&task=view&id=5&Itemid=6, accessed 10 June 2009.

WEC (2007) '2007 Survey of Energy Resources', World Energy Council, online resources: http://ny.whlib.ac.cn/pdf/Survey_of_Energy_Resources_2007.pdf, accessed 20 July 2009.

Whittaker, T.J.T, Beattie, W., Folley, M., Boake, C., Wright, A., Osterried, M. and Heath, T. (2005), 'The Limpet Wave Power Project: The First Years of Operation', Heriot-Watt University, online resources: www.sbe.hw.ac.uk/staffprofiles/A/bdgsa/shsg/Documents/2004sem/limpet.PDF, last accessed 10 June 2009.

8 Extension of the hypothesis to low power flux density sources

8.1 Introduction

Chapter 7 showed how extension of the hypothesis from HEP to other water-based renewable energy sources could be applied to tidal, marine current and wave energy, with lower but relatively comparable power flux densities. The hypothesis might be relevant in determining impact, though the paucity of available examples precluded tests of empirical data. In this section the hypothesis is extended to the lower power flux density sources, i.e. wind, solar and biomass, with power flux densities in the range <~200 W m^{-2} to 12 kW m^{-2}, and to geothermal. The parameters proposed – power flux density, proportion of interception and energy capture, efficiency of energy conversion and number of energy conversions – are applied to these sources' impacts.

The consequences of low energy flux density are generally to necessitate larger areas of energy collector and greater numbers of them to capture equivalent energy, as discussed in Chapter 3. This can lead to greater land use, depending on the collector characteristics; though this is not always the case, e.g. roof-top solar photovoltaic. The large collector area may need to be a prominent structure, as in the case of wind energy, or power tower concentrating solar, which will have environmental implications. Apart from the collector area, there are other effects of low power flux density.

8.2 Wind energy

Wind turbines harness the energy of moving air currents caused by pressure differences whose original source is the differential heating of areas of the globe, both at global and local scales; greater at the tropics and lowest at the poles (Taylor 1996).

Moving air is a fluid, though with an average density of 1.23 kg m^{-3} at sea level (Taylor 1996) it is some 813 times less dense than freshwater. Wind turbines therefore need a much greater swept area to harness equivalent amounts of power from the kinetic energy of moving air than water turbines do from moving water.

Wind energy is an almost mature technology now, which is competitive on good sites with some of the conventional fossil fuels. There is at the time of writing over 300 GW of wind energy installed around the world (WEC 2014), with some 25 years or more of commercial experience. Over this period turbines have been getting larger and more reliable, resulting in falling generation costs. Some European countries now generate large amounts of their electricity from wind energy, such as Denmark (39 per cent) (Rasmussen 2015), Spain (21 per cent) and Germany (9.7 per cent), with the UK at ~8 per cent (DUKES 2015), with over 8.4 GW installed at the time of writing, including offshore.

The environmental impacts of wind energy have been much studied, with an extensive literature accumulated, including works by this author (Clarke 1995). In the UK particularly, resistance to wind developments on the basis of environmental impacts has been slowing the development of wind capacity and leading to moves to site more capacity offshore (SDC 2005).

The most apparent effects of wind energy are from the highly visible large structures with long, rotating blades. Landscapes can be significantly altered by windfarms, especially as the scale of turbines has increased, currently up to about 120 m. The visual impact of wind turbines has been extensively studied. Other impacts include noise, though this is receding as an effect due to improved design. Electromagnetic interference to radio transmissions of various types is also possible and can influence siting. Safety and wildlife effects have further effects on siting, as does grid connection. Such impacts have been extensively described by this author and others (Clarke 1988). The impacts of wind energy are summarised in Table 8.1.

8.2.1 Wind energy power flux density

The power flux densities of energy in the wind are cited in Table 3.1 as 22–168 W m^{-2} on average to >10 kW m^{-2} maximum (Clarke 1993), and a rating of 0.471 kW m^{-2} of wind turbine converter area was achieved

Table 8.1 Summary of the main environmental impacts of wind energy

- Visual
- Noise
- Electromagnetic
- Safety
- Wildlife
- Land use/use of space
- Incompatibilities
- Construction
- Grid
- Planning/compatibilities

Source: Clarke (1995)

(Brocklehurst 1997). Power produced by windfarms in the UK ranges from 2–4 W per square metre of land area across which windfarms are sited (MacKay 2008).

The power in the wind is described by the equation:

$$P(\text{in W}) = 1/2\rho \times A \times V^3$$

Where:
P = power in watts
ρ = pressure at sea level kg m^{-3}
A = swept area of turbine blades
V = wind speed in m s^{-1}

(Taylor 1996: 276)

It can be seen from the equation above that the power in the wind increases as a cube of the wind speed. Therefore there is a focus on higher wind speed sites in choosing locations for wind turbines, which vary greatly in available power, and hence wind energy is site-specific. Power from the wind is also proportional to the swept area and so increases in blade size result in more power.

The actual power harnessed by a wind turbine is less than the power in the wind resulting from its kinetic momentum energy because the wind behaves as an extended fluid flow, and the Betz limit of 0.59 applies (Betz 1920, 1966). Typically the overall efficiency of a wind turbine's energy performance is likely to be ~42–45 per cent at optimum wind speeds (Ozgener 2006).

8.2.2 Power flux density

Applying the key parameters of the theory, and considering the first parameter, power flux density, the consequences of low power flux density of ~77 W m^{-2} to ~10 kW m^{-2}, derived from the 5 m s^{-1} to 25 m s^{-1} approximate wind speed operating range and power equation for wind, require a large area of the flow to be intercepted in order to convert significant amounts of energy with many collectors, each sweeping a large area. This results in many large structures spread across a large area. The number of turbines and land area required for 10 per cent of UK electricity supply has been calculated at 0.3–0.5 per cent of the UK land area, with less than 1 per cent of this used for foundations and roads (BWEA 2009).

Land use

The type of land use is significant for wind energy due to incompatibilities with the categories shown in Table 8.2, all unavailable for wind development.

Table 8.2 Land categories unavailable for wind energy

- Designated landscape areas
- Designated nature reserves
- Within built-up areas
- Within military zones
- Within forested areas
- Within designated amenity areas
- Within quarries and earthworks
- Steep slopes

Source: Rand and Clarke (1991)

This is largely for reasons of visual impact, obstruction of wind flow, safety issues or physical obstruction from the large structures. However, wind energy is compatible with agriculture, either arable or grazing.

Very little land is actually occupied by wind turbines, only about 0.3–2 per cent of the perimeter area of the farm (BWEA 2009; Clarke 1988), including the access tracks. Due to the need to space out turbines to avoid each other's wakes (BWEA 2009), windfarms occupy relatively large areas, although farming can continue around the turbines. It could therefore be said that wind turbines are spread *across* landscapes; occupation does not preclude all other uses.

The use of larger turbines which intercept more of the wind flow at greater heights can reduce the number of turbines required for a given output. Correspondingly greater power flux densities can be achieved than for smaller turbines.

Larger turbines, being more visible from afar, will, however, have a more extensive zone of visibility, and greater zones of visual impact (ZVI) (Engstrom and Pershagen 1980). Additionally, larger turbines will be more intrusive in relation to the scale of other features in the landscape, and the largest machines with total heights over 100 m can 'belittle' hills of similar heights. Higher power flux densities for wind energy can result in reductions of direct use of land as described below, but some indirect effects on land uses may rise in proportion to scale. For example, bigger foundations will be required, though fewer will be required per unit of installed capacity.

The effective installed power capacity density for wind energy has been cited at 9 MW km^{-2}, based on six turbines of 1.5 MW each per km^2, and a wind speed of 7.0–7.5 m s^{-1} at a height of 45 m (BWEA 2000). Employing a 500 m buffer zone for a 10 per cent UK target for wind contribution to UK electricity, 2,515 × 1.5 MW turbines would be required. In the south-west region the share would be 10 per cent of the UK's 10 per cent wind target, a 0.18 per cent land area, requiring turbines to be spread across 41.9 km^2, but only occupying ~1 per cent of this (BWEA 2009). The great increase in turbine size and power, currently up to about 100–120 m height with a rating of 5 MW or more, may affect the installed power capacity density, as larger turbines intercept greater wind speeds at greater heights above the ground, but it is likely that 9–10 MW km^{-2}, may be near the limit for most sites.

The BWEA report above refers to the ETSU report reviewing the wind resource, ETSU R-99 (Brocklehurst 1997). This shows the effect of increased power flux density for wind turbines due to large turbines. In Brocklehurst's (1997) 'A Review of the UK Onshore Wind Energy Resource', an updated and revised version of previous wind resource estimates is made. This takes into account the increasing size of commercial wind turbines – at that time 600 kW and hub heights of 45 m above ground level, as compared to 300 kW and 25 m hub above ground level for the previous estimate. Due to the reduced ground friction at higher levels, air flows are significantly greater at the higher hub height due to the one-seventh power rule (Twidell and Weir 1986: 232), and the power flux density of the flow is higher. This has the effect of increasing the power resource density per km^2 capable of being exploited to 9 MW km^{-2} from 15 turbines per km^2, compared to 3.6 MW km^{-2} from nine turbines of 400 kW.

MacKay stated that wind energy across the UK has typically a power density of 2 W m^{-2}, i.e. 2 MW km^{-2}, based on a wind speed of 6 m s^{-1}, which he takes as an average output value (MacKay 2008: 35), though subsequently he has cited a range of ~2–4 W m^{-2} (see above). The mean annual wind speed of 6 m s^{-1} is higher than many inland low-lying areas of the UK, though standard wind speed height monitoring usually refers to 10 m or 50 m above ground level. Typical commercially available turbines were in the range of 1.5–2 MW and up to 3.6 MW with hub heights of 50–100 m by 2010. Prototype 5 MW turbines were developed by the first decade of the twenty-first century (Wizelius 2008), while prototypes of up to 10 MW are being developed currently. Models of boundary shear employ a one-seventh power rule, depending on ground roughness, and so greater wind speeds and hence power flux density can be found at current typically higher hub heights. Additionally, windier locations will preferentially be chosen, so although the average of 2 W m^{-2} might prove representative of the whole country, in practice windfarms may tend to have higher power flux densities. Packing densities for windfarms are assumed to be no greater than five diameters by MacKay (2008: 266), and although this is a standard value, closer spacing perpendicular to the predominant wind direction of three diameters may be possible, depending on the wind rose distribution of wind direction.

8.2.3 Proportion

Applying the second parameter, proportion of the flow intercepted and extracted, it can readily be appreciated that for wind energy in most circumstances, only a small proportion of the flow can actually be intercepted. This is in part due to the great volume and extent of the flow itself.

For example, a low-pressure wind system in the UK may extend to a height more than 3,000 m. Some weather systems can occasionally reach greater heights than the upper troposphere (NASA 2008). The lower layer of the atmosphere, the troposphere, extends to ~10,000-15,000 m (Blackmore and Barrett 2003) and most weather systems occur in this layer. Up to about 1,000 m, friction with the ground influences air circulation considerably, and boundary layer processes apply. Surface roughness and relief features greatly affect flows near the surface. As pressure decreases with height, the mass per cubic metre of air at 5,500 m will only be about half that at sea level, which will halve the energy output from turbines for a specific wind speed.

There are many different types of airflow; in the UK most of the stronger winds are associated with low-pressure weather systems associated with the Gulf Stream, or on the outer edges of high-pressure systems. However, other parts of the world may experience different types of wind and air currents. For instance, the differential warming rates of land and sea produce onshore and offshore breezes. Mountainous areas or deserts can produce air flows of a more limited nature than the large weather systems rising to high altitudes. In some cases valleys conduct significant air flows. In certain locations the proportion intercepted by wind turbines may be significant, where the flows are of limited overall height and extent.

The size of turbines affects the proportion of the flow intercepted, for example the increase in exploitable resource cited by Brocklehurst (1997) per square kilometre is in part due to larger turbines of 45 m diameter rather than 30 m diameter machines, an increase of ~707 m^2 to 1590 m^2 per turbine, an increase of 2.25 times. So a greater proportion of the flow is intercepted.

However, although the largest currently available wind turbines exceed ~100 m total height, only a very small proportion of the total flow of a low-pressure weather system is actually intercepted and this is only a fraction of the flow available at 100 m above ground level. Nevertheless, this flow represents the current technical potential, i.e. the maximum attainable with current technology blade diameter limit and total height limit of turbines. So the measure of proportion intercepted could be taken only from the flow in the lowest ~120 m from ground level.

The fraction of the flow intercepted could be ascertained by:

- optimum maximum turbine density;
- optimum maximum commercial turbine scale;
- overall coefficient of power including the Betz limit: 0.593, e.g. 0.40 (Twidell and Weir 1986);

as a proportion of the energy in the flow available in the first 120 m above ground level, taking into account the coefficient of friction with the ground.

It can be seen that the proportion of the current technical maximum flow intercepted is much greater than of the total flow. For instance, in Figure 8.1 A–B represents the technical (current) local flow and C–D

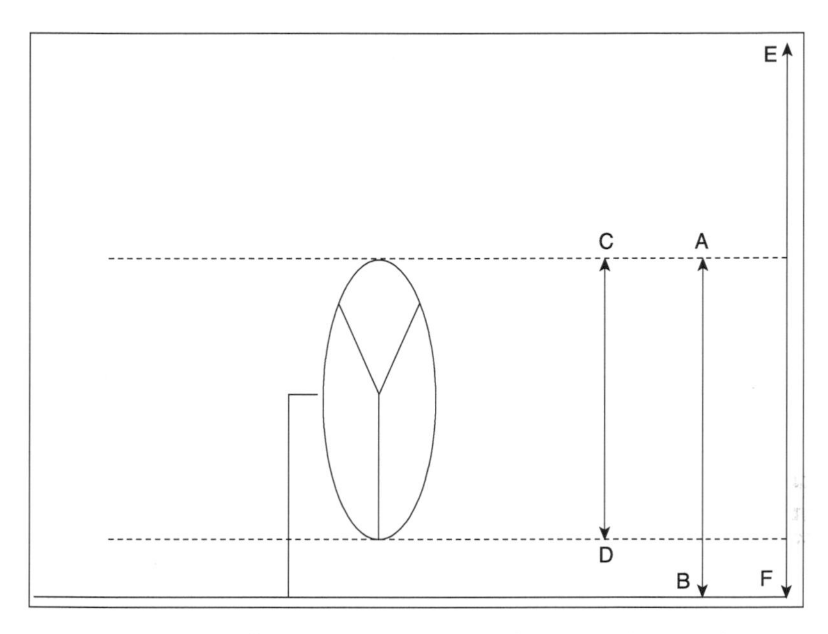

Figure 8.1 Diagram of wind turbine and ratio of intercepted and total air current flow.

represents the flow intercepted, with CD:AB the ratio intercepted. E–F represents the total height of moving air current.

The ratio AB:CD is much greater than CD to EF, where EF extends to the height of the total air current flow.

Application of this concept of proportion of flow to wind energy concerns not so much impacts from physically abstracting energy, i.e. reductions of functions of the flow, as this will usually be a relatively small proportion, but impacts from the conversion technology, i.e. turbines' unwanted by-products such as visibility and noise. In this case it is the proportion of the flow accessible within technical limits, intercepted, that will be significant. In practice the full technical potential wind resource is rarely exploited even for relatively small areas. The translation of these concepts into ones that could be of use for planning concerns the determination of the local flow.

Local flow

For any given area and scale, the wind flow can be determined; e.g. a house or vantage point, village or settlement has a breadth in relation to the prevailing winds. If the height of the flow is assumed to extend to ~120–150 m, i.e. the current approximate technical height limit to flow interception, then the total volume of the flow can be calculated by multiplying the frontal area (of the area in question), breadth × height. From any house, vantage point or settlement, the proportion of the local flow intercepted by turbines can then be calculated. What this could provide is the relative proportion of the flow as experienced at that point.

Resource density proportion intercepted

Another method of considering the concept of proportion of the flow intercepted is to consider the fraction of the resource intercepted compared to the maximum technical resource exploitation possible. This could be done for different scales. The appropriate scale would depend on the purpose; e.g. measuring impact at a single building or a settlement or other land use area.

8.2.4 Concentration of conversion

The hypothesis holds that changes in the power flux density of the flow as a result of interception could cause impact. However, this may apply more to the functions of the flow rather than 'conversion by-products', such as turbulence or noise, and more to defined flows with distinct boundaries, e.g. river water flows. Concentration of the energy conversion, after interception, in one location, e.g. with a larger turbine, might be said to concentrate impacts at that location, aside from the higher proportion of the flow being intercepted there. However, entraining the flow to achieve higher power flux density up stream of the converter could alter the flow's functions. To what extent impacts will be concentrated at a single large turbine depends to some extent on the type of technology employed; however, it will also be intrinsic to the larger turbine. Offsetting this is the fact that fewer turbines will be required and larger turbines convert more wind energy per square metre due to intercepting higher wind speeds at a greater height with lower ground friction (Twidell and Weir 1986).

By-products from conversion at each turbine cause effects such as noise. So the impact for a windfarm could be the sum total of each turbine's noise and visual and other effects, though such impacts diminish with distance and do not necessarily scale proportionately. Although higher intensity impacts may be found locally in the vicinity of a larger converter, spread across the broad area of the flow, impacts may be reduced by larger converters, since fewer conversion processes (with their by-products) are made per unit of energy.

8.2.5 Application of the parameters/hypothesis to size of turbines and impact

Addressing the issue of size of turbines and impact, the parameters can be applied to wind energy as follows.

Power flux density for Regional Target Planning Guidance (BWEA 2001) has been assumed as 9 MW km^{-2}, based on 6 × 1.5 MW turbines, with an average wind speed of 7.5 m s^{-1} at 45 m height, i.e. a 259 W m^{-2} power density of the wind flow. This is a greater power flux density than if smaller turbines had been employed, as wind speed increases with height above ground on approximately a one-seventh power rule, due to lower ground friction. This might imply a slightly greater impact, but the difference in power flux density is small, and possibly not significant.

Proportion of the flow intercepted is greater for a larger turbine, e.g. 45 m diameter turbines with a greater surface area than a 30 m diameter turbine. This could imply an increased impact, though depending on what is taken as the total flow.

Efficiency of conversion device can be greater for a larger diameter turbine, e.g. in terms of visual impact, due to the higher power per square metre of blade area. Additionally, the inevitable losses of conversion are reduced with fewer converters required per unit of electricity, implying a reduced impact. The number of energy conversion processes is therefore reduced by having fewer turbines overall. This gives rise to an overall lower level of by-products such as noise, again implying a reduced impact. Finally, there is no increase in the power flux density of the flow for wind energy, in that the flow is not concentrated by ducting or concentrators, as compared to HEP schemes. As can be seen, no values are attributed here to the increased or reduced impact. If it is assumed that visual impact is the main effect, and larger turbines are to be employed, there will be gains from needing to use fewer, but some losses in their vicinity as they could be more intrusive at close quarters.

Hypothesis as related to impacts from wind energy

The much lower density of air at sea level (1.23 kg m^{-3}) compared to freshwater (1,000 kg m^{-3}) means that the environmental effects of exploiting wind energy do not affect geomorphic processes in the same way that moving water does. Although in desert areas wind blown sand is an agent of erosion and geomorphic processes, the total proportion of energy extracted from the flow by wind energy development is likely to be small, with correspondingly small effects on geomorphic processes. Unless very large amounts of wind energy are deployed, the interception rate will tend to be too low to cause climatic effects due to extraction of energy (see discussion in Chapter 2 on macro effects). This suggests that the environmental impacts of wind energy may affect mainly the human environment or anthroposphere, and the biosphere in terms of wildlife, mainly birds, and the plant habitat at the site to some extent, depending on site characteristics.

8.3 Solar energy

Solar energy is one of the most diffuse of the direct renewable energy flows, with a power flux density of 1.367 kW at a mean earth distance from the sun and normal to the solar beam, known as the solar constant (WEC 2007).

At the earth's surface, the maximum intensity is 1 kW m^{-2} (Twidell and Weir 1986) due to atmospheric absorption and the mean is about 170 W m^{-2} when averaged over the surface of the earth and over the year (WEC 2007) or 188 W m^{-2} (Smil 2015), as the sun's angle changes diurnally and seasonally. The low power flux density requires a large collector area to harness significant quantities of energy.

There are various different types of solar energy conversion technology: flat-plate thermal, photovoltaic (PV) (electric), parabolic concentrator thermal (electric), power tower thermal (electric), and dish concentrator thermal (electric). Solar energy installations and applications are widespread across the globe (Boyle 1996), particularly in the case of flat-plate thermal solar heaters for water (or air) heating, and photovoltaic panels. Such applications have been growing fast, with over 234 GW of solar thermal by early 2012 (Mauthner and Weiss, 2013) and more than 134 GW of photovoltaic capacity employed worldwide by early 2014 (IEA 2014b) and 177 GW by the end of 2014 (IEA 2015), which produced ~200 TWh in 2015, about 1 per cent of global electricity. Solar thermal electric applications have also been expanding in recent years, though the technology dates from the 1970s.

8.3.1 Solar impacts

These different types of solar converter have varying impacts and varying power flux densities. The author has reviewed the impacts of solar energy (Clarke 1995), including hypothetical applications such as orbiting solar satellite PV with microwave transmission of the power to earth. It is outside the scope of this book to repeat this review of impacts here. A summary is included in Table 8.3.

Land use

Although a large area is required to harness solar energy due to its diffuse nature and thus the land requirement would appear to be large, in practice

Table 8.3 Summary of general potential environmental impacts from solar energy

- Land use/use of space
- Incompatibilities
- Planning
- Manufacture
- Decommissioning
- Some noise (for concentrating thermal conversion)
- Visual
- Safety
- Glare
- Wildlife

Source: Clarke (1995)

this only applies to non-building-based collector systems, such as solar power stations. Solar electric conversion technologies such as photovoltaics in solar farms or solar thermal-electric systems, such as parabolic concentrating collector installations (e.g. Luz in California) and power tower systems, examples of which exist in California and Spain (US DoE 2008), are used in systems independent of buildings.

Most direct solar energy application is expected to be in decentralised applications on the surface of the buildings (roofs or walls), where the energy will mostly be used on-site. That this can be done is due to the thin, flat collectors of photovoltaic systems and solar thermal systems, which in turn derives from the low power flux density and the technology employed. This technology involves direct conversion of solar radiation to electricity for photovoltaics and direct conversion to heat in flat-plate thermal collectors, without the need for further conversions into other energy forms, or further concentration, i.e. increases in power flux density.

Hill *et al.* (ETSU 1994) calculated that the south-facing roof area of buildings in the UK was sufficient at that time to generate 120 per cent of the UK electricity demand then prevailing, even at the relatively low efficiencies of solar PV. Hence there could, in practice, be little land use associated with solar photovoltaics application on a large scale in the UK if building-mounted collection were adopted, despite the low power flux density.

While this would appear to contradict the general tendency for lower power flux density sources to require more land area, this feature can be related to power flux density in the following manner. The solar radiation energy flows of PV and solar thermal (water heaters) are harnessed at either roughly the same power flux density of the incident solar radiation at that site, or at slightly higher concentration through built-in optical concentration (e.g. Fresnel lenses), and this low power flux density form is then compatible with ambient environmental conditions on building surfaces. These types of technology produce few hazards, noise or emissions, as the power ranges in terms of voltage, current or temperature and pressure are relatively low per square metre.

This points to the significance of the type of technology used for harnessing renewable energy as a determinant in the overall resulting environmental impact. It suggests that the direct conversion of the low power flux density by these technologies (solar PV and solar water heating) could result in lower impacts, in terms of land requirement, and possibly in other respects too. Large-scale focused solar (e.g. CSP, CPV or solar parks) may be a different issue, as is discussed later.

Resource variation

In a rather more obvious respect, power flux density variations might be seen to relate to environmental impact resulting from differences in the overall resource distribution across the globe. The total annual solar radiation flux to earth varies by a factor of about two (WEC 2004: 297) or perhaps three. The highest values experienced in, e.g. the Red Sea and Sahara Desert, at ~300 W m^{-2} or 250-273 Wm^{-2} (Smil 2015) annual mean, vary to some of the lowest values in, e.g. Northern European regions such as the UK at ~100 W m^{-2} or ~1,000 kWh m^{-2} per annum, or parts of Scandinavia and the Arctic, down to about 800 kWh m^{-2} (WEC 2007: 383). This variation results from differences in latitude, affecting the angle of the sun's rays to the surface, and in climate, cloud cover in particular. Southern parts of Europe have almost twice the solar resource as the regions with the lowest amounts (Boyle 1996).

Areas with higher solar resource totals, for the same area of collector, could harness roughly twice the energy or more, or alternatively require half the collector area. In practice, since solar converters currently depend largely on the availability of direct solar radiation, and higher latitude areas tend to have more indirect radiation, differences in the effective resource tend to be greater than suggested by the 2–3 average world solar resource variation. If the solar collector area is converted into impact in terms of manufacture and disposal, as well as use of (some) space, the impact per kWh could be half or less. It could then be stated that solar could produce twice or more the impact in terms of land use per kWh in lower solar flux regions.

This can be translated into power flux density in terms of the reduced maximum kW per square metre experienced in the UK compared to the tropics; around <80 per cent of the 1 kW per square metre maximum with the sun directly overhead, and the reduced time spans in which that higher solar flux occurs, with long periods of low power flux density. For example, in winter the highest values at the solstice will be in the region of 20 per cent or less.

However, this relationship to power flux density may be somewhat tenuous, as it concerns the distribution and frequency of the solar flux peaks and higher values, rather than the values of the maxima, which only vary by ~20 per cent. The impacts of manufacturing and disposal of collectors and the ancillary system components, although they are the main ones cited for PV and solar thermal, are considered to be relatively minor (ExternE 1995).

8.3.2 Concentrating solar thermal technologies

The relatively low power flux densities of solar energy, even in desert areas, have led to the development of concentrating technologies for thermal energy conversion which relies on high temperature difference for efficient thermodynamic conversion. The main types are parabolic mirrors concentrating solar radiation on a pipe in series or, alternatively, tracking mirrors reflecting onto a central 'power tower' point to achieve the desired concentration (Boyle 1996).

The Nevada Solar 1 Parabolic CSP plant in Nevada, USA, near Boulder City has collectors occupying 1.42 km^2 to provide continuous power of

22 MW, or a maximum of 64 MW (US DoE 2008). The actual 'solar field' mirror area is cited as 357,200 m^2. At average output, the power density would be 22 MW/1.42 km^2 = 15.5 W m^{-2}, and a maximum of 45 W m^{-2} for the whole plant.

The peak power fluxes for the collector area could be expected to be close to 1 kW m^{-2}, with the losses accounted for by the conversions from electromagnetic radiation to heat in the form of the synthetic oil, then the conversion of this heat through a heat exchanger to raise steam, followed by thermodynamic conversion to rotational shaft power by the turbine, and then finally to electricity.

The maximum theoretical efficiency ε for thermodynamic conversion is given by:

$$\varepsilon = \frac{T_{in} - T_{out}}{T_{in}}$$

Where:
T is in Kelvin
T_{in} = inflow temperature
T_{out} = outflow temperature

(Boyle 1996: 75)

At the design temperature of 371 °C, field inlet temperature is 350 °C and field outlet temperature is 395 °C (NREL 2008). The concentration ratio is 71 and the optical efficiency is cited as 0.77 (NREL 2008). The turbine steam pressure is 102 bar and 371 °C inlet temperature; the reheat pressure is 17.5 bar; steam outlet temperature is not given. The steam turbine generator's gross output is 75 MWe, and the net output to the grid network is 70 MWe. Plant parasitic energy requirements, such as for pumping the circulant and steam condenser, plus reflector tracking, account for the remaining 5 MWe. Annual electricity production is estimated to be 140–150 GWh.

Therefore the average power produced would be 145 GWh/8,760 = 16.6 MW. The average solar irradiation is 6.75 to 8–8.25 kWh m^{-2} per day in this area of Nevada (NREL 2008). An average solar radiation power flux density per 24-hour day is then 7.5/24 = 313 W m^{-2}.

This produces an overall efficiency of:

357,200 m^2 × 313 W = 111.8 MW average solar energy input
16.6/111.8 = 0.14 overall efficiency, or 14 per cent

Power flux density, proportion and efficiency

In terms of the hypothesis, the power flux density for Nevada Solar 1 is a maximum of almost 1 kW m^2, and an average of 313 W m^{-2}. The concentration is then 71 times, giving a maximum power flux density of 54 kW m^{-2} at the heated pipe, taking into account the optical efficiency, and an average there of 17 kW m^{-2}.

The proportion of the solar radiation intercepted is unknown, but likely to be ~35 per cent, taking into account reflector spacing, though the energy extracted will be 0.77 of this (optical efficiency), together with any other losses.

The efficiency of conversion is 0.14 overall, being optically 0.77. Hang *et al.* (2007) cite overall annual efficiencies of 10–15 per cent for parabolic trough systems, which is in agreement with the figures above. They cite land use at 6–8 m^2/MWh p.a. (or 16.3 W m^{-2}, though this value may not be a whole plant one) and a thermal cycle efficiency of 30–40 per cent (Hang *et al.* 2007: 258). Elsewhere, greater land use is cited, e.g. Ong *et al.* (2013) cite 15.78 m^2/MWh, or 3.9 acres per GWh p.a. in an NREL report.

The number of conversion processes is five or six, including the reflection and concentration. First solar radiation is reflected; second, it is concentrated 71 times; third is conversion to heating the circulant; fourth is conversion to steam; steam to mechanical shaft power; and finally to electricity at the generator.

Impacts as related to parameters

According to the hypothesis, land use should be inversely proportional to power flux density of the ambient energy flow. Unlike wind energy, incompatibilities with other land uses such as farming at the plant site and the need to avoid over-shading require an almost complete occupation of the site.

However, since the required insolation rates of >6.75 kWh m^{-2} per day (281 W m^{-2} average) are largely only found in desert areas, the need for large areas is less likely to be a problem; there are few, if any, competing human land uses due to the dry conditions, which together with low latitudes result in the high insolation rates.

Visual impacts are relatively high, though since population densities are generally very low in such regions there could be lower sensitivity to this. The optical efficiency of 0.77 cited will cause some reflection and so glare. Some noise may result from the steam turbine and circulating pumps. Safety concerns are unlikely to be an issue except in relation to the extensive heated circulant pipes, steam boiler and turbines and condenser. However, this may mainly affect only on-site staff and possibly wildlife.

Impacts of solar CSP and opposition

Although wildlife effects may be assumed to be small due to the low density of life in such arid regions, there have been protests against CSP solar developments. In California a sudden increase in CSP solar scheme applications from none in 2006 to 125 in 2008, for 70 GW total capacity resulting from the 20 per cent renewables target and high energy prices, led to protests which resulted in a moratorium on further applications (*New York Times* 2008). This was itself subsequently reversed. Impacts cited by the opponents (mainly local residents) include the large area occupied, water consumption for cleaning mirrors, chemicals used to clear undergrowth below the panels/reflectors, as well as loss of habitat for endangered species such as the ground squirrel, the desert tortoise and the burrowing owl. In the Nevada Desert one of the prime endangered species and habitats is that of the desert tortoise (Beacon Solar 2008). Opponents to CSP schemes propose use of dispersed roof-mounted PV as an alternative.

On the one hand, the high ratio of coverage required for concentrating solar power plants and their large area results in an intrusive development, which would appear to be incompatible with any other land uses in the same area. On the other hand, there may even be positive effects, such as the shade created below solar reflectors, where dew and soil moisture is more likely to be retained, as suggested by a study of microclimatic responses to solar collectors or mirrors in a Sonoran desert (Smith 1981). Although shading levels and panel/mirror coverage in this case was high, resulting in a 91 per cent midday reduction of the midday solar irradiance, there was a net gain in photosynthesis due to increased plant cover from reduced soil temperatures, increased moisture retention in the dry season, and reduced wind. However, the practice of using herbicide to clear the ground cover below panels and reflectors, to avoid accumulation of dust and organic detritus, will negate any gains in photosynthetic activity.

CSP power tower

The second type of CSP solar energy technology uses tracking mirrors to reflect sunlight and concentrate it on a heat exchanger on a tower, which then heats synthetic oil to high temperature to raise steam, and so turn a turbine generator unit (Boyle 1996). Solar thermal concentrating technology for electricity generation requires solar resources of over 5 kWh m^{-2} per day, which occur in latitudes below 40°. It is generally limited to sunny desert areas, e.g. in the south-west of the USA. In the early 1980s several prototype systems, e.g. at Barstow California, were constructed in which a field of reflecting tracking mirrors – heliostats – reflected the sun's rays onto a boiler at the top of a tower. Temperatures of over 500 °C were reached in the 10 MW system, and synthetic oil or molten rock salt used to carry the heat away. Since the 1980s, solar CSP power tower systems' total capacity has gradually increased, and is at the time of writing experiencing increasing growth, due in part to the ability to incorporate thermal storage and generate for several hours after sunset. Sicily, Spain, France and the Crimea also have such systems. Improvements in technology and storage techniques, as well as higher energy prices, have stimulated further interest.

Environmental impacts of CSP power tower solar plant

POWER FLUX DENSITY

In terms of the hypothesis, the ambient power flux density of about 1 kW m^{-2} maximum and average of about 170–188 W m^{-2} at the earth's surface (WEC 2013, Smil 2015) is increased by the reflecting heliostat mirrors concentrating the solar radiation in order to increase the temperatures at the converter and thereby the efficiency of thermal conversion, which depends on ΔT. Concentration factors of 300–1,000 are cited by Muller-Steinhagen and Trieb (2004).

The temperature of the power tower receiver area would be above 500 °C, and appear to glow white when in operation. A certain amount of re-radiation would occur.

EFFICIENCY

Annual average solar efficiency of 10 per cent for demonstrated CSP power tower plant and 15–25 per cent for projected schemes has been quoted (Muller-Steinhagen and Trieb, 2004), with peak solar efficiency of 20 per cent demonstrated and 35 per cent projected. As an example of CSP solar thermal technology, the 392 MW Ivanpah scheme in California, at the time of writing the largest of its type in the world, can be described. The solar irradiance resource for the Ivanpah scheme is cited as 7.4 kWh m^{-2} per day (NREL 2014), or an average of about 308 W m^{-2} with a total of 2,717 kWh m^{-2} per year (Clean Energy Action 2013). The expected output is 1,079 GWh per year. The total heliostat area is 2,347,144 m^2, so to find the efficiency:

$$1{,}079 \times 10^9/(2{,}717 \times 10^3 \times 2{,}347{,}144) = 0.169$$

On this basis, the efficiency of 0.169, or about 17 per cent, would, if achieved, be a relatively high value indicating relatively efficient land use.

PROPORTION

The approximate proportion of energy intercepted can be found by dividing the reflector area by the area they are spread across. Thus, for the 392 MW installed capacity of Ivanpah CSP facility, with three power towers, the total reflector area is given as 2,347,144 m^2 (Clean Energy Action 2013) and the total area of the site is given as 14,164,150 m^2 (*CSP Today*, 2014a). So approximately one-sixth, or 16.6 per cent, of the solar radiation flow is intercepted over the whole site, though interception may be higher in the vicinity of the power towers.

NUMBER OF ENERGY CONVERSIONS

The number of conversions is seven: first, reflection; second, concentration; third, heating of the receiver; fourth, conversion to the heated circulant; fifth, conversion of heat to steam; sixth, conversion to shaft drive via a turbine; and lastly, to electrical energy by turning a generator. The first three of these conversions processes are 'uncontained', i.e. they are carried out on the radiation flow in the open.

LAND USE

Power tower solar CSP is cited as having a land use of 8–12 m^2 MWh^{-1} per year by (Muller-Steinhagen and Trieb, 2004). The Ivanpah CSP power tower scheme covers a total of 1,600 ha, so at a forecast output of 1,079 GWh p.a., this would result in a land use of 14.8 m^2 MWh^{-1} p.a., slightly greater but roughly in accordance with the figures given by Muller-Steinhagen and Trieb.

DISCUSSION OF CSP POWER TOWER IMPACTS

Visual impacts: as well as the land area required for the heliostat reflectors, the 137 m (450 ft) high towers and ancillary turbine and generator housing with steam condensing and electrical equipment housing will constitute visual impacts.

According to the hypothesis proposed here, the number of energy conversions, together with the much greater change in power flux density of CSP power tower schemes, is likely to result in more impacts per unit than from CSP Parabolic Trough schemes. Is this borne out in practice? Although there are as yet relatively few CSP Power Tower developments to furnish evidence, or CSP Parabolic Trough schemes for firm comparisons to be drawn, it appears that this may be the case. In particular, the concentration of the solar radiation into higher power flux densities appears to result in an increased 'glare' effect, for example to pilots and aircraft (Font 2014), and there are some reports of birds being affected when flying through the more intense solar radiation field, with apparently some scorching effects being observed (Kraemer 2014). Font notes that pilots from airports 25 and 40 miles away at Las Vegas International Airport have made complaints about glare since start-up of the Ivanpah CSP plant (Font 2014). This issue is, however, quite well known in the CSP industry and provision for it was made in environmental impact statements. Apparently, the problem arises during start-up operations when some of the heliostats are directed away from the tower, to enable gradual heat build up. Heliostats in standby mode need to be directed away from the sky. Aircraft pilots need to be warned of the locations of CSP power tower stations, and possible glare hazards. Measurements of glare and light levels need to be made and be within agreed limits. Mitigation of this issue could be amenable to good planning and

Table 8.4 Summary of potential impacts from CSP solar power

- Land use/use of space
- Incompatibilities
- Planning
- Noise
- Visual
- Water demand
- Safety
- Glare
- Wildlife

operation practices, but there has not yet been extensive experience with such plants. At Ivanpah, the issue has been addressed by altering the heliostat stand-by algorithms, with some success compared to unaltered units (*CSP Today* 2014b). On the issue of bird impacts, Kraemer states that 'power towers also create areas of solar flux (basically concentrated sunlight) near the tower that have singed the feathers of some birds flying within proximity of the towers at Ivanpah during its recent commissioning and startup'. Apparently the area within 50 m of the tower is the main area of concern, apart from preventing collisions with the heliostats. This is understandable since temperatures of above 500 °C for the outlet circulant from the power tower are reached, and surface temperatures will be higher still. There have been bird mortalities showing signs of singed feathers (Kraemer 2014). The strategy is to deter birds from these areas, e.g. by using bird calls or low-frequency sounds, and monitoring is being undertaken by the developers in communication with the US Fish and Wildlife Service (USFWS).

Other wildlife affected by CSP solar power towers has included desert tortoises, requiring fencing and resulting in a temporary suspension in 2011, followed by mitigation measures.

Concentrating solar power technologies clearly can result in impacts deriving from the power flux density, e.g. in land use, the interception rate, proportion of flow intercepted and extracted, as well as the concentration – in terms of glare – and number and efficiency of conversion processes. This points to the relevance of the parameters used in the hypothesis.

8.3.3 Photovoltaic solar

Photovoltaic solar energy conversion uses semi-conductor materials to allow photons to release electrons, giving rise to a voltage difference which can be used to create a current (Boyle 1996). The semi-conductor materials consist of p–n junctions, where the doped p material contains an excess of electrons, and the doped n material has a deficiency of electrons. Silicon solar cells use high-purity silicon doped with small amounts of boron in the p layer and phosphorous in the n layer. Photons have an energy of ~2 eV and electromagnetic wave particle radiation is a highly organised low-entropy form of energy. Mono-crystalline silicon solar PV cells had attained an efficiency of about 16 per cent (Boyle 1996: 101) by the mid-1990s, though up to 22 per cent has been claimed (Sunpower Corp 2008) and 25.6 per cent by 2014 (Green *et al.* 2014). A typical cell produces about 1.25 W, with a voltage of about 0.5 V and a current of about 2.5 A (Boyle 1996). Higher conversion efficiencies can be achieved by using other materials, e.g. gallium arsenide, by multi-junction cells which can intercept more frequencies of light, and by concentration lenses. Efficiencies of over 40 per cent have been achieved, though at greater expense. Typically, cells of about 20 per cent efficiency have been the commercial limit, though efforts are being made to raise this and efficiency is gradually rising.

Generally, solar cells are assembled into large, flat panels to provide useful amounts of power. Such panels, having no moving parts and requiring only electrical connections and secure mountings, can be installed on building roofs or walls, or alternatively on frames on the ground. In higher latitudes the panel will need to be tilted to be normal to the solar beam – 30° in the UK is normal as a compromise between seasonal variations.

Solar PV will only incur extra land use when placed in solar array fields, as has been done, for example, in Germany at the Bavaria Mühlhausen Solarpark of 10.08 MW capacity (TDW 2004) and many further examples since, both in the UK and elsewhere (STA 2013).

PV solar park power stations in terms of the hypothesis

The solar irradiance energy density at the Bavaria Solarpark at Mühlhausen is ~1,350 kWh m^{-2} p.a. (Suri *et al.* 2006), or an average power flux density of 154 W m^{-2}. Here, solar PV panels with an area of 250,000 m^2 are used in the Solarpark, which is part of the Bavaria Solarpark, occupying a total area of 400,000 m^2 (Sunpower Corp 2008). Output was expected to be 6.8 GWh p.a. from Mühlhausen, and the output from the whole site is 10.85 GWh p.a., giving an average output of 0.776 MW for Mühlhausen, and 1.238 MW for the whole site, resulting in a land use of 36.87 km^2/TWh p.a. for the whole site.

Therefore, the overall average efficiency of energy capture for the total flow is approximately

$$1.238 \text{ MW}/(154 \text{ W} \times 250{,}000) = 3.2 \text{ per cent}$$

The flow interception rate would then be 250,000/400,000 = 62.5 per cent. If a panel angle of 30° is assumed, then the horizontal area occupied by the panels is reduced by ~13 per cent, so the actual horizontal area occupied by the panels would be 216,506 m^2.

A typical interception rate would be about 33 per cent for the 400,000 m^2 collector/110 ha total area PV solar park, e.g. Brandis/Bolanden near Leipzig (Juwi 2008).

A feasibility study by the ACT government in the Australian Capital Territory for a solar power station to supply 10,000 homes (ACT 2008) identified solar PV as more efficient in land use requirement than solar concentrated thermal (SCT), at 75 ha of land for 57 MW (76 W m^{-2}) installed capacity as opposed to 120 ha (47.5 W m^{-2}) for the SCT plant.

This latter example illustrates how power flux density can be used to compare and assess land use impact of different technologies, showing the value of a common parameter that can be applied to compare impacts of different schemes, together with the factors of efficiency, interception and energy extraction ratio, as well as a number of conversions.

More recently, wider experience of solar parks has led to reports of impacts related to: PAR levels (photosynthetic active radiation), wind speed effects, soil temperature and soil moisture, as well as effects on wildlife, in particular birds and invertebrates (Brooks 2014; van Rooyen and Froneman 2013).

The issue of PAR levels identified in the Rampisham Down Solar Park planning document (Brooks 2014) illustrates an example of the environmental impact parameter 'proportion of flow intercepted' in operation. In this instance, concern was raised over the level of shading of the vegetation below the solar panels, and a study of mitigation methods was carried out involving translucent panels, lowered pitch angles for the PV panels and separations between panels to allow light to penetrate. These measures resulted in a 15 per cent greater PAR. Other environmental data collection showed that a small reduction in wind speed below the panels occurred, together with a small reduction in soil temperatures (16 per cent), and an increase in soil moisture (47 per cent), with variable rates of PAR. However, no vegetation was 100 per cent shaded below the translucent panels and the authors state that significant levels of PAR reached even the most shaded areas for periods of at least ten minutes. In the context of the extremely exposed growing conditions on the site, the authors conclude that no harm will occur to the habitat from the solar park. These results echo those of Smith (1981), investigating micro-climate effects of solar PV in a Sonoran desert in Mexico. Impacts depend in part on the characteristics of individual sites.

The parameter 'efficiency of conversion' and resulting by-products can be illustrated with regard to bird impacts from large solar PV parks in the drier parts of the USA, such as California, where birds can apparently mistake distant solar PV parks for water due to the reflectivity and polarisation of light of the panels (C. Clarke 2013; Upton and Climate Central 2014).

The parameter 'proportion intercepted' can also be illustrated by larger-scale PV developments, e.g. the 550 MW PV Desert Sunlight Solar Farm being built near Eagle Mountain (C. Clarke 2013), where the size of the development, occupying 16 km^2, may be a factor in the response of birds to the 'lake effect' from reflection. It has been pointed out that the Pacific Flyway, the bird migration route along the US west coast in California, much of it in desert with its oases and stop-over points, such as the Colorado

River and the Salton Sea, now has solar PV projects which look much like lakes (C. Clarke 2013). This is thought to have led to bird mortalities. The Bureau of Land Management and the US Fish and Wildlife Service agencies are reported to be working on minimising future mortality of rare birds such as the Yuma clapper.

There are compelling economic reasons for developing large, utility-scale solar PV parks rather than residential-scale developments, since the panel purchase, installation, substation and connection, as well as project design, cost significantly less per unit generated. Costs cited range from a median of 9.15¢ per kWh for utility scale, as opposed to a median of 15.8¢ per kWh for residential scale (Tsuchida *et al.* 2015). Such large solar parks as the 550 MW Desert Sunlight Eagle Mountain scheme, occupying 16 km^2 or about 6.2 square miles, would not be applicable to UK or most European conditions.

As the scale of utility PV solar plants in desert areas increases, there will inevitably be issues relating to the parameter 'proportion of flow intercepted' in relation to the assumed flow cross-section and its overall environmental impact. This echoes the notion of 'local flow' as applied to wind energy (see above), when considering the more diffuse flows of wind and solar, which tend to be more akin to a field effect rather than a distinct defined flow such as that of rivers. Such issues have been discussed in Chapters 2 and 3, as well as for other sources.

8.4 Biomass

The case of biomass renewable energy is more complex than some of the other sources, since it is in effect converted solar energy stored in a chemical form. But its use as a fuel does enable determination of a power flux density figure based on output per unit land use requirement. Moreover, it is both a low power flux density source, derived from sunlight and, as the carbon and carbohydrate compounds and materials themselves flow through the natural environment, a more concentrated higher power flux density source.

Furthermore, since biomass sources involve carbon compounds, carbohydrates (CH_2O compounds), the building blocks of life, there are additional environmental issues such as biodiversity, as well as competition with food production to consider in their use. Moreover the storage capability of biomass energy sources introduces still further complications. Despite this complexity, the environmental effects of biomass sources can still be assessed in terms of the hypothesis described in this book.

Biomass energy sources

Biomass energy refers to the use of dead plant and animal matter for energy (Hall 1991). Biomass is derived from a large number of diverse sources, as shown in Table 8.5.

Table 8.5 List of biomass energy sources

- Urban domestic wastes
- Urban industrial wastes
- Food processing wastes
- Sewage wastes
- Forestry wastes
- Energy forestry
- Fuel crops (e.g. for biodiesel)
- Landfill gas
- Energy coppicing (SRC)
- Agricultural wastes

Source: adapted from Ramage and Scurlock (1996).

Biomass energy can be divided into two parts: production of the fuel material; and conversion into useful energy. The first part is the growing of plant matter, or collection of wastes, and the second can include the conversion to fuels with a higher energy density (either by volume kW m^{-3} or by kW per unit mass). A fuel is defined as a store of energy (Boyle 1996).

Biomass energy sources are the traditional supplies of energy used by humans ever since the use of fire was discovered. Such sources are still important in global total primary energy supply, estimated at 10 per cent (IEA 2014a) or more, since much of traditional use is in the informal sector. Traditional use is the burning of wood and plant or animal matter for cooking, lighting and heating. Although this source has declined in importance with the increasing use of fossil fuels and other energy sources, the modern use of biofuels is expanding in many countries with the development of more efficient and clean technologies for production and conversion. In the UK, for example, ~50 per cent of the renewable electricity supplied was derived from biomass sources in 2008 (DUKES 2008) and still sizable amounts more recently: 35 per cent of renewable electricity and 69 per cent of total primary renewable energy supply in 2014 (DUKES 2015).

8.4.1 Environmental impacts of biomass energy

Because biomass energy is composed of biomatter, the substance of living matter, its use as a fuel has considerable environmental implications. Production issues concern the land area required and the substitution of natural vegetation by the biomass crop, which will have ecological and biodiversity consequences, depending on the former crop or plant cover. Visual impact can be a concern from stands of crops. Wildlife can be affected due to habitat changes. Soil quality can be an issue depending on the rate of harvesting, and water and fertiliser requirements are important considerations. Biomass crops can have incompatibilities with some other land uses; for example, short rotation coppicing (SRC) may preclude simultaneous wind energy use due to increased surface roughness. Land use incompatibilities are

Table 8.6 Potential environmental impacts from biomass energy

Production	*Conversion*
Visual	Visual
Biodiversity reduction	Noise
Wildlife changes	
Soil effects	
Water requirements	
Fertilizer run-off	
Land-use	Land-use
Incompatibilities	Incompatibilities
Competition with food production	Emissions: gas, liquid, solid
iLUC (indirect land use change)	
Degree of CO_2 neutrality, carbon and energy balances	
Waste processing	

broadly similar to those of arable farming use. Biomass fuel crops compete for land use with food production and in recent times this has become a more contentious issue (Srinivasan 2008). In addition, biomass energy use is rarely entirely CO_2 neutral; that is, cultivation and harvesting as well as fertiliser inputs all require an energy input, usually in the form of carbon (fossil) fuels. The ratio of carbon saved (compared to traditional fossil fuels) can vary very widely from a fairly high to low percentage, or even be negative, with biofuel crops such as ethanol produced from maize resulting in savings of only ~33 per cent, compared to ethanol produced from sugar cane, which can produce 71 per cent savings (Gallagher 2008: 26).

Conversion of biomass into fuels and energy constitutes a very diverse topic with many varieties of processes and technologies. This author has reviewed many of these at length in previous publications (especially Clarke 1995, 2000), as have other authors (Hall 1991; Ramage and Scurlock 1996). It is beyond the scope of this book to consider all the different means of biomass production and conversion. Here, the general principles of applying the hypothesis to some selected examples will suffice. A summary of the environmental impacts of biomass energy is shown in Table 8.6.

Biomass power flux density

Biomass energy sources have the lowest overall power flux, ranging from a maximum of 1 kW m^{-2}, or about <800 W m^{-2} solar flux in, for example, the UK, to just 16 W m^{-2} actually absorbed at the leaf surface (Twidell and Weir 1986: 261), with much less available as biomatter after the plant has used the energy required for respiration (Ramage and Scurlock 1996). This can be seen from Figure 3.3.

The reason for this very low power density flux is that solar light radiation, already the most diffuse of the primary renewable energy flows, is then converted by photosynthesis by plants at low efficiency, and only a proportion of this energy is actually used for plant growth, so forming biomass materials for use as a fuel. The overall efficiency has been estimated to be as low as <~0.5 per cent of the original solar input in the UK (Ramage and Scurlock 1996: 154).

Power flux density and land area required for biomass energy

As a consequence of very low power flux density, biomass energy primary production requires very large amounts of land area (see Figure 3.4; Gagnon *et al.* 2002: fig. 4).

Therefore the production of biomass energy can be expected to have a large environmental impact in terms of land use, directly related to the low power flux density of the source production. Biomass land area productivity varies widely across the world with biome conditions, with as little as 1 t ha^{-1} p.a. biomass produced and as much as 30 t ha^{-1}. Short rotation willow coppicing can produce ~10 t ha^{-1} p.a. in the UK (Ramage and Scurlock 1996), which provides 150 GJ p.a. or 0.48 W m^{-2}, but 1.44 W m^{-2} can be achieved in optimum circumstances elsewhere.

Conversion of the biomass materials to more concentrated forms results in much higher power flux density (in terms of carbon material flows). Table 8.7 shows that the energy density for some typical biomass materials used for fuels is between one-quarter and half (or lower) that of fossil fuels. Fossil fuels are, of course, derived from biomass stock, compressed and chemically

Table 8.7 Average energy density of biomass and fossil fuel stock

Fuel	*Energy density*	
	GJ t^{-1}	*GJ m^{-3}*
Wood	15	10
Paper	17	9
Dung	16	4
Straw	14	1.4
Sugar cane	14	10
Domestic refuse (as collected)	9	1.5
Commercial wastes (UK average)	16	N/A
Grass	4	3
Oil	42	34
Coal	28	50
Natural gas	55	0.04

Source: adapted from Ramage and Scurlock (1996).

processed by geological processes, i.e. converted into the more energy-dense forms that currently provide ~80 per cent of human usage of energy worldwide (IEA 2014a).

Low energy density biomass materials are inconvenient to handle and transport, as well as being prone to decay (Ramage and Scurlock 1996). The low-value materials are uneconomic to transport and utilise in solid forms such as logs or grass, and therefore are converted to higher-energy density forms such as the more familiar fuels, e.g. alcohols and methane. The conversion processes can then have environmental implications.

Carbon flow and other cycles

Carbon flows are also important to fundamental life processes, such as the cycling of soil nutrients and the organic carbon compounds which are cycled by fungi and saprophytic organisms (Nasholm and Ekblad 1998), as well as the storage of carbon in the soil. Carbon flows through soils are to be understood as comprising both inorganic and organic carbon, with the inorganic carbon in solution being much more readily available, and the organic CH_2O compounds having greater permanence.

The soil stores about twice the carbon that the atmosphere does (IPCC 2007), and the importance of carbon stores in soil, as a sink for carbon sequestration, has been acknowledged by many commentators (IPCC 2007). Abstracting from these carbon flows in order to extract energy could result in greater cycling of the carbon via the atmosphere.

Figure 8.2 shows a simple biomass combustion cycle which assumes that all of the carbon is cycled via the atmosphere. One of the major impacts of using fossil fuels has been the rise in atmospheric carbon from ~280 parts per million (ppm) in 1800 to 365 ppm in 2003 (Blackmore and Barrett 2003), leading to average global warming of at least 0.5 °C so far (Blackmore and Barrett 2003), and now over 400 ppm, with accelerating increases (NASA 2013; IPCC 2013). This results in changing climates, with warming, sea-level rise, melting of ice caps and glaciers, as well as greater incidence of more extreme weather events. Therefore, further cycling of carbon via the atmosphere needs to be undertaken only in a sustainable, controlled manner. That means ensuring that carbon neutrality or near neutrality is achieved. The measuring of net carbon flows from the soil carbon stores and other stores is required.

In practice it is difficult to accurately measure soil carbon due to the dynamic nature of the flows as inorganic solutions and organic compounds, much of which may be in flux, is processed by microorganisms, bacteria and fungi at any one time. Figure 8.3 shows a more developed cycle for natural plant carbon, nitrogen, water and minerals cycle, with cycling of organic compounds via the soil and fungi or micorrhizal organisms on plant roots (Nasholm and Ekblad 1998). The importance of knowing the size of the

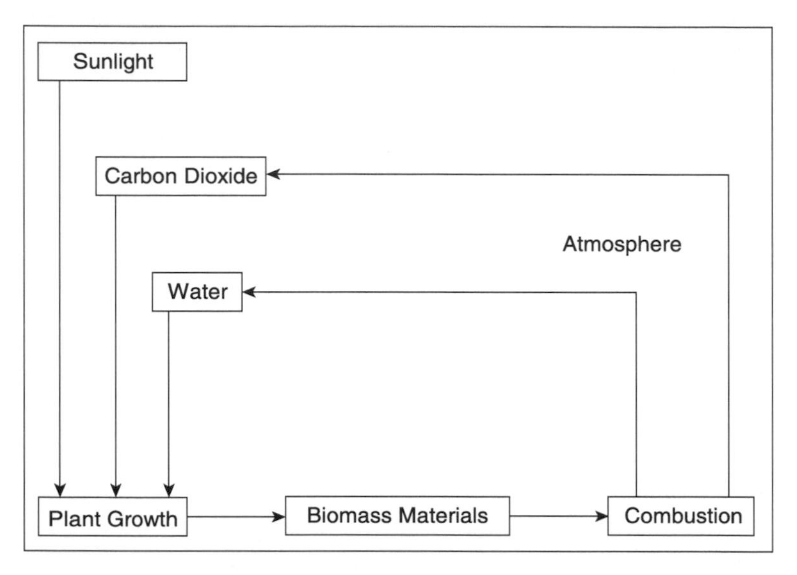

Figure 8.2 Simple biomass combustion cycle.

Source: adapted from Clarke (2000).

store, as well as the magnitude of the flows, is evident, despite the difficulties in obtaining accurate soil carbon content figures.

The difficulty of taking into account the time lags involved in carbon compound stores both in long-lived plants such as forest trees and the soil will be apparent.

Conversion of biomass fuels

There is a wide diversity of methods for biomass energy conversion. The different conversion methods are:

- direct combustion, to produce heat for direct use in an external combustion engine;
- thermochemical processing such as pyrolysis, gasification and liquefaction;
- biological processing, such as anaerobic digestion to produce CH_4, and fermentation to produce alcohols such as ethanol C_2H_5OH (Ramage and Scurlock 1996).

8.4.2 Biomass energy in terms of the hypothesis parameters

In terms of the hypothesis:

- the power flux density (available to the plant at the primary production stage) is very low at 16 W m^{-2} at the leaf, or less <~0.5 W m^{-2} as the

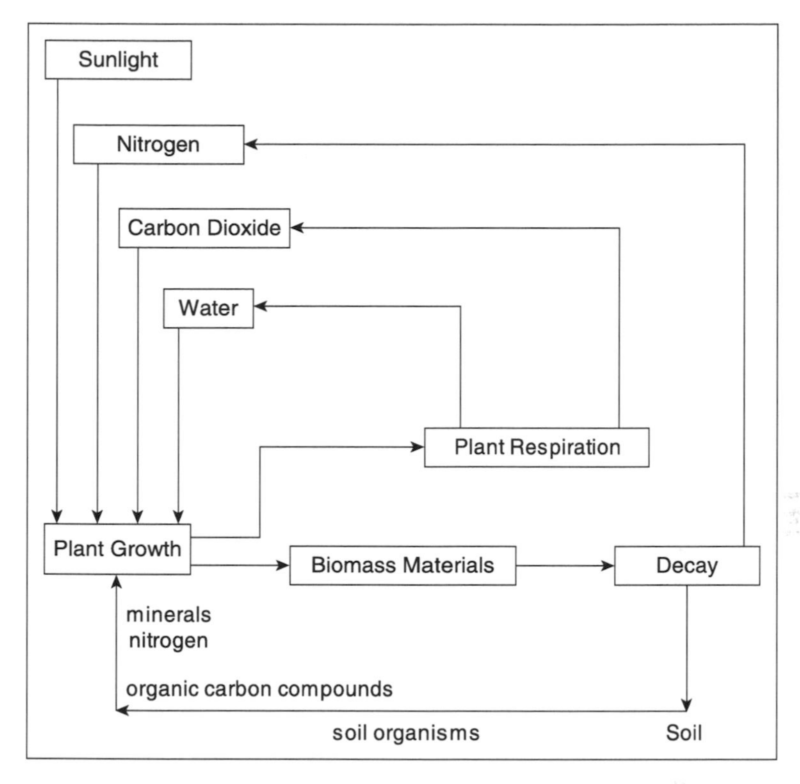

Figure 8.3 More developed biomass cycle, taking account of soil flows of organic carbon.

Source: adapted from Clarke (2000).

average in UK conditions, greater at lower latitudes, depending on the biome (climatic conditions) in question. This, as stated, leads to high land use per unit of energy obtained.

The proportion of the energy flow that is extracted needs to be no more than the plants and land area can replace on a long-term basis, while not depleting existing stores of organic carbon in the soil, to be sustainable.

The efficiency of the initial energy conversion from solar radiation to organic compounds by plants, i.e. photosynthesis and plant growth, is very low at ~2–<0.5 per cent (Tan *et al.* 2009; Ramage and Scurlock 1996: 145). Elsewhere, different figures for efficiency are cited, depending on what is actually taken into account, and the source. Twidell and Weir (1986) cite approximate photosynthetic efficiencies from data reported for many different circumstances. These range from 0.1 per cent for the whole earth, to 1 per cent for a forest annual general average, 2 per cent for grassland tropical or well-managed temperate, 3 per cent for closely planted cereal crops under good farming practice in growing season only, to 10 per cent in laboratory conditions with enhanced CO_2, optimised temperature and lighting, and

30 per cent theoretical maximum for initial photosynthetic process, with filtered light, controlled conditions and non-inclusion of plant respiration. Plants only use a proportion of the wavelength spectrum, around 45 per cent, peaking mainly in the blue and red visible part at ~0.4 μm and ~0.68 μm wavelengths, rather than the green, as measured by oxygen production, though this occurs across almost the whole spectrum (Twidell and Weir 1986).

Da Rosa (2012: 577) points out that the optimum observed values of the efficiency of photosynthesis is at a light power density of 300–400 W m^{-2}, measured by CO_2 uptake. As he states, this is well below the 1,000 W m^{-2} maximum solar power flux density at the earth's surface, and plants are optimised for the average light conditions they experience, with leaves shading each other. The requirement to use and conserve water for transpiration, as well as moderate temperatures to about 25–35 °C for optimum photosynthesis processes, are further reasons for the optimum power flux density being lower than the maximum available. Different types of plants have different efficiencies depending on the conditions for which they are adapted; C3-type plants can be more efficient than C4-type plants at high CO_2 concentrations, though C4 plants are more efficient at higher light power densities and in conditions of restricted supply of water (Da Rosa 2012).

Utilising biomass fuels for energy by combustion limits energy conversion efficiencies to those determined by the Carnot cycle and the laws of thermodynamics, so that efficiencies rarely exceed 50 per cent and are typically lower, at ~30 per cent or less. Harvesting, processing, transport and conversion involve further efficiency losses.

The number of conversions involved will be relatively high since, first, there is conversion of solar radiation to chemical potential energy by the plant (although this is a natural process and therefore could be considered to have positive impacts). Second, there may be conversion of the biomass fuel stock to a more energy-dense form, such as an alcohol or gas, or at the least harvesting involving some forms of processing. Third, conversion from potential chemical energy to heat energy or rotational shaft energy is performed by combustion in a heat engine; and fourth, to electricity in a generator.

These multiple conversions can each lead to further impacts, as efficiency losses at each stage lead to waste energy forming unwanted by-products, such as gaseous chemical compounds of unburnt hydrocarbons, e.g. carbon monoxide, particulates, nitric oxides or other residues.

As an example of the application of some of the hypothesis principles, the DECC report on the use of woody biomass from North America for electricity generation in the UK (Stephenson and MacKay 2014) can be cited. The low power flux density of the boreal forests, affecting growth rates and hence land area required, was contrasted with more southerly sourced wood fuel in terms of impact. The proportion of the energy flow (i.e. carbon fixed by photosynthesis by the tree) that is extracted for fuel is reflected in the rate of harvesting, and to what extent whole stems are used rather than thinnings, branches and sawdust, identified as a significant impact factor in the report. In addition, the diversion of the wood from its functions in nature, i.e. rotting wood cycling carbon and organic compounds via the soil with accompanying temporal storage, to atmospheric cycling, from combustion as a fuel, can be likened to impact from diversion of the other renewable energy flows functions' in nature. The efficiency of conversion is reflected in the energy balances involved in drying, pelletising and transporting the fuel. Unsurprisingly, the energy requirement of drying, and long distances involved in transportation from North America, can outweigh much of the fuel energy content available. Additionally, the concentration of biomass (energy) flows by means of long-distance transport from a wide radius to large biomass energy conversion stations illustrates the impact parameter of changes in power flux density. The number of conversions required, and the changes in energy density of the fuel, is reflected in pelletising and drying processes. Together with the low energy conversion efficiency of ~35 per cent in conventional UK thermal power stations, the overall carbon emissions savings can be nullified or even negative. The use of stem wood, i.e. whole-tree use, is such an important factor as it concerns the carbon store in the trunk and proportion extracted. The carbon stored in the trunk, in turn, is a reflection of the time lag of that part of the biomass store, and the proportion of the flow that is extracted – i.e. both a reservoir which causes a delay in the overall energy flux, which can in turn reflect the overall power flux density, and also the proportion of the flow that is extracted.

Biomass energy conclusions

Applying the hypothesis could help identify how impacts can arise from the use of biomass energy sources. The low power flux densities lead to large use of land per unit of energy, which can compete with food cultivation, and also the need to convert the fuel stock so that useful high energy density and economic forms can be employed. The proportion of the energy flow that is extracted appears to be crucial to the long-term performance of soils and also to the net carbon neutrality. Conversion efficiency is low and

Table 8.8 Summary of potential impacts from biomass energy production

- Land use/use of space
- Carbon neutrality and carbon balance
- Planning/compatibilities
- Visual
- Biodiversity effects
- Wildlife (can be positive)
- Soil effects: compaction from cultivation
- Carbon depletion
- Water requirement
- Fertiliser

multiple conversions are required – all of which can lead to further impacts. However, biomass energy flows are an integral part of the natural environment, and there are many waste streams from human food production and economic activity that can be effectively utilised. The primary production of biofuels for fuels needs the greatest scrutiny to ensure that overall gains are worthwhile, both in energy and carbon terms, as well as in the other mineral cycles and other impacts. Using biomass energy sources can produce environmental gains, e.g. for wildlife, for carbon sequestration in soils or through treatment of wastes, provided that the overall processes are carefully identified, assessed and managed. Nevertheless, biomass energy utilisation also has the potential to be unsustainable and environmentally damaging and so requires thorough assessment.

8.5 Geothermal energy

Geothermal energy is the use of the decay heat from radioactive isotopes, such as $uranium_{238}$, $thorium_{232}$ and $potassium_{40}$, and heat from the formation of the earth in the earth's crust and core (Brown 1996). It is one of the older renewable energy sources, dating from about the turn of the twentieth century (Brown 1996). It produced 67.25 TWh p.a. globally (GEA 2010), about 0.31 per cent of world electricity in 2010 (IEA 2012), mainly in volcanic or plate margin areas. Two types of geothermal energy exist potentially: wet aquifer (current geothermal energy) and hot dry rock (HDR), also known as enhanced or engineered geothermal systems (EGS), which has only been proven in prototypes at the time of writing. While wet aquifer geothermal energy is a relatively limited resource, HDR potentially has much greater total heat capacity. Geothermal energy is only partly renewable in that it represents a store of heat, albeit one which is replenished, but not over the lifetime of a development, rather over tens or hundreds of years (Brown 1996).

Approximately 12 GW of geothermal electricity generation capacity is installed worldwide (GEA 2014), plus ~15 GW thermal in use, at the time of writing. Enhanced geothermal systems are being developed which use improved drilling technology, e.g. in Germany and in the UK.

8.5.1 Environmental impacts of geothermal energy

This author has reviewed impacts of geothermal energy usage (Clarke 1995). Table 8.9 offers a summary of the impacts.

It is not proposed here to go into great detail applying the hypothesis to geothermal energy, as this is outside the scope of this study, but the method of applying the hypothesis is briefly outlined.

In terms of the hypothesis, the power flux density of the average background heat radiated from the earth's surface is very low at 0.06 W m^{-2} (Brown 1996: 358). However, the total flow amounts to 10^{21} J p.a. which is ~21 times current world total primary energy use, so the resource is

Table 8.9 Summary of potential environmental impacts from geothermal energy

- Land use/use of space
- Incompatibilities
- Planning
- Noise
- Visual
- Safety
- Wildlife
- Gaseous emissions
- Liquid emissions
- Subsidence possible
- Minor earth movements and minor seismic events

Source: adapted from Clarke (1995).

potentially large. Despite the low average power flux density, it is concentrated at certain points to an average of ~300 mW m^{-2} in the form of aquifers due to suitable geological conditions. The conditions required are permeable rock strata supplying a water source, and a cap of non-permeable rock, insulating rocks and a suitable energy gradient per depth (Brown 1996). Such aquifers can be natural features, breaking to the surface as geysers or hot springs. At such points, high power flux densities of, for example, tens or hundreds of watts per square metre can be found, but they are relatively rare.

In terms of the function in nature of this energy flow, it forms part of the geological processes and lithosphere evolution of the planet. The mantle convection current flows cause rocks to gradually cool and change chemically in so doing, at least near the surface. Mantle convection currents form a vital part of the earth's rock cycle mechanism, with the important function of long-term carbon sequestration through burying of calcium carbonate to form limestone beds. Extracting some this heat may not be significant environmentally, provided that it is the pre-existing flow which is exploited. However, exploratory boreholes are drilled to find new resources of wet aquifers, and possibly in the future for HDR wells, which could accelerate the rate of cooling that would have occurred naturally. A possible result of this is the minor earthquakes and seismic events which often accompany geothermal exploitation, though such events are often common in areas near tectonic plate boundaries (Brown 1996). If much larger energy extraction were undertaken, there could conceivably be effects on the total amount of heat available for convection currents.

8.5.2 Geothermal energy in terms of the hypothesis

In terms of the hypothesis, the increase in power flux density over that which would have occurred naturally, produced by, e.g. HDR or wet aquifer extension, could have the potential to result in environmental impact, in that changes might be accelerated. A question that could be asked is whether the greater cooling rate will cause changes such as contractions of the rocks which result in significantly increased seismic events. Some accelerated rock reservoir cooling could possibly occur almost without impact on the surface, provided that the hot rock reservoirs are not cooled too rapidly or extensively, depending on heat conductivity of the surrounding rocks and recharge rates. However, this topic must remain largely speculative here, though it is not outside the bounds of feasibility.

The proportion of the flow intercepted, as a parameter, could be applied to geothermal and HDR sources, although data are lacking in this study.

The efficiency of conversion – as related to impacts – could also be a relevant parameter for geothermal. Some of the lower temperature and enthalpy (total heat content of the system) geothermal developments have lower conversion efficiencies due to the Carnot Cycle limitations of thermal energy conversion from lower temperature resources. This can result in the requirement for more plant equipment, such as condensers, piping per unit of electricity produced (Brown 1996).

The number of energy conversions is four. First, heat energy is transferred to hot steam or water under pressure, which for the lower temperature and enthalpy developments may require 'flashing over' the hot pressurised water to steam, the heat being passed to a heat exchanger. Hot dry steam of a quality to turn turbines directly is rare, except in the highest temperature geothermal regions (Brown 1996). Second, the steam is converted to rotational shaft power; and third, to electrical form in the generator. A fourth stage is required, re-injecting condensed steam into the borehole, preventing the aquifer running dry, together with re-injecting the highly corrosive and potentially polluting minerals, which are in solution, dissolved salts and chemicals such as CO_2, HCO_3, H_2S (Brown 1996).

It is tentatively suggested therefore that the hypothesis parameters for evaluating the impacts of renewable energy sources could be applied meaningfully to geothermal energy too.

8.6 Conclusion

The parameters identified in the hypothesis could be applied to the lower power flux density sources of renewable energy to help identify, analyse and explain environmental impacts. The lower power flux densities can result in greater land use per unit of energy, as can be seen especially in the case of solar power stations and biomass production. Higher power flux densities, as achieved by larger wind turbines and windier locations, or areas of high solar energy, e.g. above 1600 kWh m^{-2} p.a., or biomass production from more productive biomes and species, conversely appear to use less land area. However, the sensitivities of other impacts to variations in power flux density are perhaps lower, given the smaller variations in power flux densities, e.g. solar energy fluxes vary by a factor of only about 2.5–3 (Boyle 1996) across the globe.

Low power flux density can result in compatibility with human habitation, since solar PV can be incorporated into dwellings. Where the energy conversion technologies are direct, i.e. involving only one conversion, and with little further concentration, e.g. solar photovoltaic, impacts could be minimised, with the result that panels can be placed on roofs of buildings which are already occupying land. Solar PV generation need then require little extra land. Biomass sources from waste streams can also result in minimal further land use, too.

The concept of proportion of energy flow extracted is harder to define for the low power flux sources, although still apparently valid, because the flow itself is much less defined than those of, e.g. water flowing in a river. The air currents of winds occur as broad flows, while those of solar are universal across the globe, though varying by latitude, and diurnally, seasonally and climatically. Biomass energy depends crucially on the productivity of biomes, particularly the availability of water and temperature. As a result of this difficulty in defining these flows, the notion of local flows has been proposed, and so the proportion of energy flux intercepted and extracted could then be determined locally. This would appear to apply to wind or central solar power stations or even biomass.

The nature of wind energy generally, as broad and high air currents, together with the Betz limit and the need to disperse turbines, can result in only a small proportion of the energy being extracted, unless the flow is of limited cross-section and volume. The proportion of solar energy extractable is limited by conversion technologies that cannot yet achieve very much more than ~30–40 per cent maximum efficiencies, with 15–20 per cent being more common for commercial panels, although the proportion intercepted is often about ~33 per cent at the plant, for central solar plants, whether PV or concentrating solar thermal type.

The parameter 'efficiency of conversion' can be applied to these renewable energy sources to help determine their environmental impacts, particularly in the case of biomass, where very low overall energy conversion efficiency is found from solar to biomass energy end use, e.g. about <0.5 per cent (Boyle 1996) to 2 per cent (Tan *et al.* 2009). As yet, solar electric conversion rarely achieves efficiencies of 30 per cent and, generally about 15–20 per cent or less is more likely for CSP or PV. Wind energy conversion efficiency is limited by the theoretical Betz limit, as well as turbine coefficient – which would need to be optimised for the range of wind speeds found at the site.

The number of energy conversions as a parameter of impact could be relevant for solar, biomass and wind, though maybe to a lesser degree. For biomass, where a large number of conversions are carried out this parameter could also be informative of impacts.

These examples serve to show that the hypothesis can be meaningfully employed to analyse impacts from lower power flux density sources.

Figure 8.4 Solar concentrating solar power tower plant by permission of Abengoa Solar Solúcar Solar Power Plant. Sanlúcar la Mayor, Sevilla, Spain (technology owned by Abengoa Solar, S.A.) ©Abengoa Solar, S.A. 2014. All rights reserved.

Figure 8.5 Solar CSP Iberdrola showing land use, by permission of Abengoa Solar Solúcar Solar Power Plant. Sanlúcar la Mayor, Sevilla, Spain (technology owned by Abengoa Solar, S.A.) ©Abengoa Solar, S.A. 2014. All rights reserved.

Figure 8.6 Solar photovoltaic park, Oxfordshire, UK; Westmill Solar Co-op aerial view. Copyright Neil Maw, by permission of Neil Maw ©.

Figure 8.7 Solar farm Gloucestershire, UK. Copyright Alexander Clarke ©.

Figure 8.8 Domestic solar photovoltaic installation. Copyright Alexander Clarke ©.

Figure 8.9 Clyde Wind Farm, supplied by permission of SSE. Copyright Noel Cummins ©.

Figure 8.10 Bryn Titli wind farm, Central Wales, UK. Copyright Alexander Clarke ©.

Figure 8.11 Biomass production SRC trials in Buckinghamshire. Copyright Alexander Clarke ©.

Bibliography

ACT (2008) 'Solar Power Pre-Feasibility Study', Parsons Brinkerhoff Australia pty Ltd ActewAGL and Australian Capital Territory Government, online resources: www.cmd.act.gov.au/__data/assets/pdf_file/0005/2939/Solar_Power_Plant_Pre-feasibility_study.pdf, accessed 28 November 2008.

Beacon Solar (2008) 'Beacon Solar Energy Project Certification Application Siting Cases Executive Summary', Government of California, online resources: www.energy.ca.gov/sitingcases/beacon/documents/applicant/afc/1.0%20Executive%20Summary.pdf, accessed November 2008.

Betz, A. (1920) 'Das Maximum der theoretisch möglichen Ausnutzung des Windes durch Windmotoren', *Zeitschrift für das gesamte Turbinewesen*, 20 September: 307–9.

Betz, A. (1966) *Introduction to the Theory of Flow Machines*, trans. Randall, D.G., Pergamon Press, Oxford.

Blackmore, R. and Barrett, R. (2003) 'Dynamic Atmosphere' in D. Morris *et al.*, eds, *Changing Environments*, John Wiley & Sons Wiley and the Open University, Milton Keynes.

Boyle, G. (ed.) (1996) *Renewable Energy Power for a Sustainable Future*, Oxford University Press in conjunction with the Open University, Milton Keynes.

Brocklehurst, F. (1997) *A Review of the UK Onshore Wind Energy Resource*, HMSO, London.

Brooks, J. (2014) 'Environmental Data Report', Rampisham Down Light Levels Monitoring Initial Report, Community Heat & Power, online resources: http://wam.westdorset-dc.gov.uk/WAM/doc/Planning.pdf?extension=.pdf&contentType=application/pdf&id=1107232, accessed 26 July 2015.

Brown, G. (1996) 'Geothermal Energy', in Boyle, G., ed., *Renewable Energy Power for a Sustainable Future*, Oxford University Press in conjunction with the Open University, Milton Keynes.

BWEA (2000) *Planning for Wind Energy: A Guide for Regional Developers*, British Wind Energy Association, Proceedings of 2000 22nd BWEA Wind Energy Conference.

BWEA (2001) 'Regional Target Planning Guidance', British Wind Energy Association, online resources: www.bwea.com/planning/uk_planning_legislation.html, accessed 7 August 2009.

BWEA (2009) 'Wind Energy Frequently Asked Questions', British Wind Energy Association, online resources: www.bwea.com/ref/faq.html, accessed 11 June 2009.

Clarke, A. (1993) 'Comparing the Impacts of Renewables: A Preliminary Analysis', TPG Occasional Paper, Technology Policy Group 23, Open University, Milton Keynes.

Clarke, A. (1995) 'Environmental Impacts of Renewable Energy: A Literature Review,' Thesis for a Bachelor of Philosophy Degree, Technology Policy Group, Open University, Milton Keynes.

Clarke, A.D. (1988) 'Windfarm Location and Environmental Impact', Charter for Renewable Energy and Network for Alternative Technology and Technology Assessment, c/o Energy & Environment Research Unit, Open University, Milton Keynes.

Clarke, A. (2000) 'Life Cycle Assessment of Energy from Biomass Waste Sources', Energy and Environment Research Unit, Report 075, Open University, Milton Keynes.

Clarke, C. (2013), 'Water Birds Turning Up Dead at Solar Projects in the Desert', *ReWire, Redefine*, 17 July, online resources: www.kcet.org/news/redefine/rewire/

solar/water-birds-turning-up-dead-at-solar-projects-in-desert.html, accessed 26 July 2015.

Clean Energy Action (2013) 'Ivanpah Solar Electric Generating Station (ISEGS) Case Study', Clean Energy Action Project, online resources: www.cleanenergyactionproject.com/CleanEnergyActionProject/CS.Ivanpah_Solar_Electric_Generating_Station___Concentrating_Solar_Power_Case_Study.html, accessed 4 July 2014.

CSP Today (2014a) '8th Concentrated Solar Thermal Power Conference & Expo', *CSP Today*, 5–6 June, online resources: www.csptoday.com/usa/site-visit.php?utm_source=PR&utm_medium=portal&utm_content=portal&utm_campaign=PR+portal, accessed 23 June 2014.

CSP Today (2014b) 'How Ivanpah is Reducing Glint and Glare from Heliostats', *CSP Today*, 4 September, online resources: http://social.csptoday.com/technology/how-ivanpah-reducing-glint-and-glare-heliostats, accessed September 2015.

Da Rosa, A.V. (2012) *Fundamentals of Renewable Energy Processes*, 3rd edition, Elsevier, New York.

DUKES (2008) 'Digest of UK Energy Statistics', BERR Business Enterprise and Regulatory Reform Department, online resources: www.berr.gov.uk/energy/statistics/source/renewables/page18513.html, accessed 11 June 2009.

DUKES (2015) 'Digest of UK Energy Statistics', Department of Energy and Climate Change, online resources: www.gov.uk/government/statistics/renewable-sources-of-energy-chapter-6-digest-of-united-kingdom-energy-statistics-dukes, accessed 5 September 2015.

Engstrom, S. and Pershagen, B. (1980) 'Aesthetic Factors and Visual Effects of Large Scale WECS', National Swedish Board for Energy Development NE.

ETSU (1994) *An Assessment of Renewable Energy for the UK*, HMSO, London.

ExternE (1995) *Externalities of Energy*, vol. 1, European Commission, Luxembourg.

Font, V. (2014) 'Concentrating Solar Power Under Fire: Glaring Planning Oversight or Easily Remedied Issue?', *Renewable Energy World.Com*, 8 April, online resources: www.renewableenergyworld.com/rea/news/article/2014/04/concentrating-solar-power-under-fire-glaring-planning-oversight-or-easily-remedied-issue, accessed 10 July 2014.

Gagnon, L., Belanger, C. and Uchiyama, Y. (2002) 'Life Cycle Assessment of Electricity Generation Options: The Status of Research in Year 2001', *Energy Policy*, Vol. 30: 1267–78.

Gallagher, E. (2008) 'The Gallagher Review of the Indirect Effects of Biofuels Production', Renewables Fuels Agency, online resources: www.renewablefuelsagency.org/_db/_documents/Report_of_the_Gallagher_review.pdf, accessed 31 December 2008

GEA (2010) 'Geothermal Energy International Market Update', Geothermal Energy Association, online resources: www.geo-energy.org/pdf/reports/GEA_International_Market_Report_Final_May_2010.pdf.

GEA (2014) '2014 Annual US and Global Power Production Report', online resources: http://geo-znergy.org/events/2014%20Annual%20US%20&%20Global%20Geothermal%20Power%20Production%20Report%20Final.pdf, accessed 3 February 2016.

Green, M., *et al.* (2014) 'Solar Efficiency Tables Version 44', *Progress in Photovoltaics: Research and Applications*, Vol. 22: 701–10.

Hall, D.O. (1991) 'Biomass Energy', *Energy Policy*, October: 711–37.

Hang, Q., Zhoa, J., Xiao, Y. and Junkui, C. (2007) 'Prospect of Concentrating Solar Power in China: The Sustainable Future', *Renewable & Sustainable Energy Reviews*, Vol. 12: 2505–14.

IEA (2012) 'Key World Energy Statistics', International Energy Agency, online resources: www.iea.org/publications/freepublications/publication/KeyWorld2014.pdf, accessed 20 May 2015

IEA (2014a) 'Key World Energy Statistics', International Energy Agency, online resources: www.iea.org/publications/freepublications/publication/KeyWorld2014.pdf, accessed 20 May 2015.

IEA (2014b) 'PVPS Report Snapshot of Global PV 1992–2013' International Energy Agency, online resources: www.iea-pvps.org/fileadmin/dam/public/report/statistics/PVPS_report_-_A_Snapshot_of_Global_PV_-_1992–2013_-_final_3.pdf, accessed 21 June 2014.

IEA (2015) 'Snapshot of Global PV Markets', online resources: www.iea-pvps.org/fileadmin/dam/public/report/technical/PVPS_report_-_A_Snapshot_of_Global_PV_-_1992–2014.pdf, accessed 7 September 2015.

IPCC (2007) 'Climate Change 2007: Synthesis Report Summary for Policymakers', Intergovernmental Panel on Climate Change, WMO, UNEP, online resources: www.ipcc.ch/pdf/assessment-report/ar4/syr/ar4_syr_spm.pdf, accessed 20 July 2009.

IPCC (2013) 'Climate Change 2013: The Physical Science Basis,' 5th assessment cycle, online resources: www.climatechange2013.org, accessed 3 February 2016.

Joyce, S., (2015) 'On Denmark's Road to Renewable Power', *Inside Energy*, 24 June, online resources: http://insideenergy.org/2015/06/24/on-denmarks-road-to-renewable-power, accessed 27 January 2016.

Juwi (2008) 'World's Largest Solar Power Plant Being Built in E Germany', press release, online resources: www.tu.no/multimedia/archive/00036/Presentasjon_av_Wald_36754a.pdf, accessed November 2008.

Kraemer, S. (2014) 'How to Protect Birds from CSP Towers', *CSP Today*, 6 June, online resources: http://social.csptoday.com/technology/how-protect-birds-csp-towers?utm_source=http%3a%2f%2fuk.csptoday.com%2ffc_csp_pvlz%2f&utm_medium=email&utm_campaign=CSP+eBrief+9+June&utm_term=Will+India%e2%80%99s+pro-business+leadership+help+gr, accessed 10 July 2014.

MacKay, D. (2008) 'Technical Chapters to Sustainable Energy', online resources, http://www.withouthotair.com/cft.pdf, accessed 13 May 2016.

Mauthner, F. and Weiss, W. (2013) 'Solar Heat Worldwide, Markets and Contribution to the Energy Supply 2011', IEA, SHC, online resources: www.iea-shc.org/solar-heat-worldwide, accessed 21 June 2014.

Muller-Steinhagen, H. and Trieb, F. (2004) 'Concentrating Solar Power: A Review of the Technology', Institute of Technical Thermodynamics, German Aerospace Centre, online resources: www.dlr.de/Portaldata/41/Resources/dokumente/institut/system/publications/Concentrating_Solar_Power_Part_1.pdf, accessed 24 June 2014.

NASA (2008) 'RST Section 14: The Water Planet', NASA, online resources: http://rst.gsfc.nasa.gov/Sect14/Sect14_1c.html, accessed 10 November 2008.

NASA (2013) 'Global Climate Change, Vital Signs for the Planet', National NASA, online resources: http://climate.nasa.gov/evidence, accessed 1 August 2014.

Nasholm, T. and Ekblad, A. (1998) 'Boreal Forest Plants Take Up Organic Nitrogen', *Nature*, Vol. 392: 30.

New York Times (2008) 'Solar Projects Draw New Opposition', *New York Times*, 23 September 2008, online resources: www.nytimes.com/2008/09/24/business/businessspecial2/24shrike.html?pagewanted=2&_r=1&ref=businessspecial2, accessed November 2008.

NREL (2008) 'Overview of Concentrating Solar Power Technologies', National Renewable Energy Laboratory, online resources: www.energy.ca.gov/reti/environmental_com/2008-04-04_CSP_OVERVIEW.PDF, accessed November 2008.

NREL (2014) Ivanpah solar Electric Generating System, online resources: http://www.nrel.gov/csp/solarpaces/project_detail.cfm/projectID=62, accessed 12 May 2016.

Ong, S., *et al.* (2013) 'Land Use Requirements for Solar Plants in the United States', National Renewable Energy Laboratory, online resources: www.nrel.gov/docs/fy13osti/56290.pdf, accessed 19 September 2015.

Ozgener, O. (2006) 'A Small Wind Turbine Application and Its Performance Analysis', *Energy Conversion and Management*, Vol. 47 No. 11–12: 1326–37.

Ramage, J. and Scurlock, J. (1996) 'Biomass', in Boyle, G., ed., *Renewable Energy Power for a Sustainable Future*, Oxford University Press in conjunction with the Open University, Milton Keynes.

Rand, M. and Clarke, A. (1991) 'The Environmental and Community Impacts of Wind Energy in the UK', *Wind Engineering*, February.

SDC (2005) 'Wind Power in the UK', Sustainable Development Commission, online resources: www.sd-commission.org.uk/publications.php?id=234, accessed 18 August 2009.

Smil, V. (2015) Power Density: A Key to Understanding Energy Sources and Uses, Cambridge, MA: MIT Press.

Smith, D.S. (1981) 'Microenvironmental Responses of a Sonoran Desert Ecosystem to a Simulated Solar Collector Facility', PhD Thesis, University of California.

Srinivasan, S. (2008) 'The Food v Fuel Debate: A Nuanced View of Incentive Structures', *Renewable Energy*, Vol. 43: 950–4.

STA (2013) 'Recent Examples of Commercial Scale PV', Orta Solar, online resources: www.solar-trade.org.uk//media/Orta%20Solar%20Case%20Studies.pdf, accessed 16 August 2014.

Stephenson, A. and MacKay, D. (2014) 'Life Cycle Emissions of Biomass Electricity in 2020: Scenarios for Assessing the Greenhouse Gas Impacts and Energy Requirements of Using N. American Woody Biomass for Electricity Generation in the UK', Department of Energy and Climate Change, 2014.

Sunpower Corp (2008) 'Bavaria Solarpark Exceeds Energy Expectations', Sunpower Corp, online resources: www.sunpowercorp.it/For-Power-Plants/~/media/Downloads/for_powerplants/spwr_bavaria_en_a4_w.ashx, accessed 28 November 2008.

Suri, M., Huld, T.A., Dunlop, E.D. and Cebecauer, T. (2006) 'Photovoltaic Electrical Potential in European Countries', European Commission Joint Research Centre, Institute for Environmental and Sustainability Renewable Energies Unit, online resources: http://re.jrc.ec.europa.eu/pvgis/cmaps/eu_opt/PVGIS-EuropeSolarPotential.pdf, accessed 28 November 2008.

Tan, K.T., Lee, K.T., Mohamed, A.R. and Batia, S. (2009) 'Palm Oil: Addressing Issues and Towards Sustainable Development', *Renewable & Sustainable Energy Reviews*, Vol. 13: 420–7.

Taylor, D. (1996) 'Wind Energy', in Boyle, G., ed., *Renewable Energy Power for a Sustainable Future*, Oxford University Press in conjunction with the Open University, Milton Keynes.

TDW (2004) 'Germany: Siemens PTD and PowerLight Corp. to Build Solarpark in Bavaria', TDW Transmission and Distribution World, online resources: http://tdworld.com/substations/power_germany_siemens_ptd, accessed 1 December 2004.

Tsuchida, B., *et al.* (2015) 'Comparative Generation Costs of Utility-Scale and Residential-Scale PV in Xcel Energy Colorado's Service Area', The Brattle Group, prepared for First Solar, online resources: www.brattle.com/news-and-knowledge/news/study-by-brattle-economists-quantifies-the-benefits-of-utility-scale-solar-pv; for the report see: http://brattle.com/system/publications/pdfs/000/005/188/original/Comparative_Generation_Costs_of_Utility-Scale_and_Residential-Scale_PV_in_Xcel_Energy_Colorado%27s_Service_Area.pdf?1436797265, accessed 27 July 2015

Twidell, J. and Weir, A. (1986) *Renewable Energy Resources*, E&FN Spon, London.

Upton, J. and Climate Central (2014) 'Solar Farms Threaten Birds', *Scientific American*, 27 August, online resources: www.scientificamerican.com/article/solar-farms-threaten-birds, accessed 26 July 2017.

US DoE (2008) 'Concentrating Solar Power-Parabolic Reflectors', United States Department of Energy, online resources: http://solareis.anl.gov/documents/docs/NREL_CSP_3.pdf, accessed November 2008.

van Rooyen, C. and Froneman, A. (2013) 'Solar Park Integration Project, Bird Impact Assessment Study', Revised Final Report, South African Heritage Resources Agency, online resources: www.sahra.org.za/sahris/sites/default/files/additionaldocs/Appendix%20H4%20-%20Avifauna%20Impact%20Assessment.pdf, accessed 26 July 2015.

WEC (2004) '2004 Survey of Energy Resources', World Energy Council, online resources: www.worldenergy.org/documents/ser2004.pdf, accessed 20 July 2009.

WEC (2007), '2004 Survey of Energy Resources', World Energy Council, online resources: http://ny.whlib.ac.cn/pdf/Survey_of_Energy_Resources_2007.pdf.

WEC (2013) 'World Energy Resources: Solar', World Energy Council, online resources: www.worldenergy.org/wp-content/uploads/2013/10/WER_2013_8_Solar_revised.pdf, accessed 1 October 2015.

WEC (2014) '2014 Marked a Record Year for Global Wind Power', online resources: www.gwec.net/global-figures/wind-energy-global-status, accessed 20 June 2014.

Wizelius, T. (2008) *Developing Wind Power Projects*, Earthscan in association with IIED, London.

9 Conclusions

9.1 Introduction

This book has addressed a hypothesis on the broad topic of environmental impact from the diverse renewable energy sources. Environmental impact has been defined as changes beyond the 'normal' rate of change, that is faster or greater in extent, or more permanent changes. The term 'environment' covers many different concepts, from physical surroundings to processes and systems, including the lithosphere as well as the biosphere, atmosphere and anthroposphere. An attempt at some more succinct definitions of subcomponents has been made here in respect of the different aspects of the environment that are affected by renewable energy sources. The concept of power flux density has been used to characterise renewable energy sources in terms of their environmental effects and impacts. Of necessity, there has been considerable simplification and generalisation in application of the concept to environmental impacts, but these can be justified.

Nevertheless, an attempt has been made to link the functions of the natural energy flows and their role in the environment to changes caused by intercepting and extracting energy by renewable energy conversion systems. While it has been proposed that this may apply to all of the different renewable energy sources, the focus of the detailed element of the study has been on hydro electric power (HEP) as a suitable case with sufficient available data to demonstrate the principle. Hence much of the study has been concerned with one renewable energy source, first considering effects at the dam and reservoir and then extending this to downstream effects on available energy and its processes. The specific processes that have been considered here have been those of energy available for sediment erosion and transport, although there are many other impacts related to energy, but more indirectly, such as effects on wildlife. Inevitably, it has only been possible to study certain, more direct processes of rivers here, and their relationship to environmental effects and impacts of HEP, even though many indirect effects such as water quality are related.

9.2 Study review and findings by chapter

9.2.1 Theory

Most renewable energy sources (apart from geothermal and tidal) are ultimately derived from the sun, at low power flux densities, less than 1 kW m^{-2} at the earth's surface, and an average of ~170–188 W m^{-2}. These energy flows are successively converted into more concentrated forms, such as wind at up to about 12 kW m^{-2}, and water flows such as wave and river flows, successively increasing the power flux density, with waterfalls or HEP schemes of more than 100 kW m^{-2}. Wave energy comes in between. Eight different flows are identified in this book: solar, biomass, HEP, tidal, marine current, wind, wave, geothermal. They can be characterised by their power flux densities.

The definition of the term 'power flux density' is the rate of work of the energy flow per unit of area. The work that natural energy flows perform in the environment can be classified as a set of roles or 'functions' in powering and maintaining natural physical and living systems. By definition, natural energy flows with higher power flux densities will have a higher rate of work per square metre. The hypothesis is that the degree of interference with such flows by extracting energy or changing its distribution will accordingly result in impact, as a result of changing the energy and power available for the natural functions. Furthermore, the intensity or power flux density of the energy flow will be a factor in the land use of renewable energy. The higher the power flux density, the less land will be required per unit of energy captured – in theory. This may seem obvious in the case of solar or wind; diffuse energy flows mean larger catchment areas are required. However, the precise interaction may be more complex.

Environmental impact is defined as changes. In addition to extraction of energy, environmental impact also results from the production of unwanted by-products, as well as perturbation of the flow. Therefore, in addition to power flux density, the following parameters have been proposed:

- proportion of energy intercepted and extracted;
- efficiency of conversion;
- number of conversions of energy form;
- changes in power flux density.

These four variables are not independent of each other, but interdependent. But power flux density may be the most significant variable since the proportion of energy captured from a particular renewable source appears to depend in part on the power flux density of the flow, as shown in Figure 3.6, and see also Figure 3.5.

The purpose of the project was to compare different renewable energy sources, technologies and schemes with one another in terms of environmental impact, defined as changes in land use terms, in terms of interference with the natural functions of each energy flow, and in terms of unwanted by-products from conversion processes.

This parametric approach to the theory of environmental impacts from renewable energy sources is considered by the author to be a novel one, which has not been found in the literature, except for previous work outlining the approach by this author.

9.2.2 The HEP test

The hypothesis was tested on a selection of HEP schemes noted for their environmental impacts. Data for a number of parameters were collected and interpreted in power flux density form for the sample. The test consisted of two parts: first, impacts at the dam and reservoir reach were tested for the parameters; and second, impacts downstream from the HEP scheme using the stream power concept as a measure of power flux density. Over 60 different spreadsheets were employed in assembling the data. Since the topic of environmental impacts is very diverse, two key impacts were chosen for testing for correlation to power flux density – sediment transport and land use.

Test results

For the first part of the test on the dam and reservoir reach, no apparent relationship between land use and power flux density derived from (hydraulic) head height was found for the sample, nor between land use and total stream power for the reservoir reach. No relationship between land use and former mean gradient was found for the sample. The results suggested that there could be an inverse correlation between the cross-sectional stream power and land use (i.e. with slope and flow rate). A power equation for CSP to land use could be implied, $y = 67566x^{-0.4219}$, with $R^2 = 0.606$, i.e. impact (land use) $I = 5E + 07(CSP)^{-1.4371}$. However, this was not statistically significant, also due to the small sample size and the nature of the relationship cannot be stated with confidence. Also, a linear correlation between water residence time and land use (slope $y = 1.7037x$) could be indicated by the sample data. That is, the land use of storage in the sample was ~0.469 days per km^2/TWh p.a., though this was not statistically confirmed. A possible linear relationship between land use and reservoir volume was also indicated, suggesting that despite differences in reservoir shape and depth, generally reservoirs of greater volume occupied more land per unit of energy generated. This implied that variations in reservoir shape and depth can be largely discounted as a factor accounting for the data. Again, though, due to the small sample size confidence in this was low. No relationship between land use and average reservoir fluid velocity was indicated. No clear relationship between land use and the production factor was apparent, though dams designed for considerable storage appeared to be the lower production factor schemes. No apparent relationship between land use and mean river flow was shown by the data. Brune sediment trap efficiency compared to head height and hypothetical power flux density (pfd) might indicate a linear correlation with a slope of $y = 0.7743x$ and $R^2 = 0.5048$, i.e. I (Impact) = 0.7743 $pfd_{hypothetical}$ (hypothetical power flux density), or with a power correlation a slope of $y = 2.9823x^{0.7264}$ and $R^2 = 0.8306$, i.e. $I = 2.9823(pfd_{hypothetical}{}^{0.7264})$, though the sample was also too small for this to be statistically significant.

The second part of the test on four rivers downstream of the HEP scheme estimated stream power values before and after the HEP scheme. Pre- and post-dam, total stream power (TSP) estimated for four rivers in the sample, from gradient profiles downstream of the major dams, showed considerable variation in decrease of TSP below HEP dams. The decreases in TSP ranged from a maximum decrease of 98.6 per cent on the Colorado River at Yuma (near the Mexico border) for the period 1933–75, to a 10.4 per cent decrease in peak monthly TSP on the Danube below the Iron Gates dam post-construction, and an almost imperceptible decrease in TSP below the Gabcikovo scheme, though neither of the latter can be stated with confidence due to the relatively short data period. For the Columbia River a peak monthly TSP reduction of 51 per cent was shown below the Grand Coulee dam, after construction of the Mica dam upstream of this, although 670 km downstream near the mouth, a reduction of only 23 per cent was seen. On the Nile, a drop in monthly peak TSP of 71 per cent below the Aswan High dam was seen after construction, which is likely to continue downstream, though long-term data were not available to confirm this. The very high TSP decreases for the Colorado are in part also due to abstraction of water flows for irrigation and water supply, with average flow rates near the mouth decreasing by 96 per cent between 1933 and 1975.

The estimation of TSP was carried out where sufficient data were available, pre- and post-dam, but there were considerable discontinuities in both the timeline and river line data sets that were available, and so this work is far from a complete modelling exercise for the river downstream of HEP schemes. Nevertheless, with the extensive data set, the author considers the validity of the application of TSP in this context to have been demonstrated.

The test compared the reduction in peak TSP with delta land area loss for three rivers which had deltas, out of the four estimated for TSP. The results were indicative of a correlation between TSP reduction and delta land area loss. The two rivers that had considerable reductions in TSP – the Nile (71 per cent TSP reduction) and the Colorado (98 per cent reduction) – caused by the HEP schemes both showed considerable delta land area loss, while the Danube (6.9–10.4 per cent TSP reduction), which had small or imperceptible TSP reductions, showed a much smaller delta land area loss. Thus those dams that could pass the flood on down the river appeared to cause little loss of sediment transport, while those that attenuated the flood significantly or completely reduced or almost terminated transport of sediment. Reduction of peak TSP (pTSPr) compared to delta land area loss impact produced a linear slope of $y = 13.564x$, with $R^2 = 0.5945$, i.e. $I = 13.564$ (pTSPr). Although inconclusive, the number of cases was too small, at only three, to be statistically significant, the author considers this might be indicative of support for the hypothesis. TSP can be interpreted as a measure of power flux density, and the reduction of sediment transport is a reduction in one of the primary functions of the natural energy flow. Thus the change in power flux density appears to be related to environmental impact for HEP.

The extent of the task on some of the world's longest rivers precluded a larger sample.

What this would appear to indicate is that the major impoundment dams, which store large volumes of water flow, can cause impacts in terms of loss of sediment transport, whereas the 'run of the river' dams whose reservoirs do not store river flow pass through the sediment due to the higher power flux density of the unattenuated flood flows. This is not stating that sedimentation occurs, but the reverse – that loss of the flood peak stream power causes *loss* of sediment pick up and transport all the way downstream. A common misconception is that sediment originates only upstream, whereas it can originate all along the bed, pick up and deposition being in dynamic balance (Petts 1984).

9.2.3 Other water-based technologies

The hypothesis was extended to other water-based renewable sources – tidal barrage, marine current and wave energy – and to the lower power flux density sources – those below 1 kW m^{-2} average, which are wind, solar, biomass and geothermal. However, no tests of data were carried out so this section was exploratory and somewhat speculative. These two categories were discussed in Chapters 7 and 8.

It was proposed that impacts may be related to the power flux density for the other water-based sources, as well as to the additional parameters employed here, since similar processes, i.e. sediment transport and deposition, were involved. Comparing tidal barrage and marine current turbines, power flux density differences, proportion of interception and energy extraction, as well as number of conversions, pointed to the relevance of the hypothesis. However, it was beyond the scope of the study to verify this with tests of data. The few existing examples of tidal barrages, marine current and wave energy schemes worldwide as yet would make it difficult to verify this.

The much higher power flux density of tidal barrages at the converter, of up to 500–628 kW m^{-2} (Severn Barrage), compared to marine current turbines' average of ~<10 kW m^{-2}, appears likely to be reflected in the lower impacts that may be experienced from marine current turbines.

The flow interception proportion of tidal barrages is ~100 per cent, but probably a maximum of ~<30 per cent for marine current turbines, although for tidal fence schemes it may be higher. The energy extraction proportion has been cited at 30 per cent for tidal barrages (Charlier 2003), for marine current turbines it would probably be lower, depending on the scheme.

The number of conversions of form of energy in the flow is only one for marine current turbines, compared to three for tidal barrages, and this could be expected to result in fewer impacts, e.g. sedimentation and tidal range. The efficiency of energy conversion could be higher for tidal barrages in terms of interception, at ~65–85 per cent as for low-head HEP, compared to the marine current turbine of ~45 per cent (see Chapter 7).

For wave energy converters, average power flux densities of ~12 kW m^{-2} for offshore wave energy converters on the west coasts of the UK and Ireland can be cited. At least 50 per cent of this would be dissipated in bottom friction near-shore and up to 80 per cent by the shoreline, representing natural functions performed, such as erosion, sediment stirring, oxygenation and beach building. Higher power flux densities could result in a greater rate of such functions, and diminution of average power flux densities through energy extraction appears likely to result in impacts such as, changes in wave height and beach shape, and related geomorphic and biological effects. Although studies of currently proposed developments such as the SW Wave Hub envisage only small changes at the shore of 2.3 per cent diminution, greater deployment densities would result in greater changes, reflected in the proportion of energy interception and extraction, converter efficiency and number of conversions. A 22–32 per cent reduction in energy levels could be expected 1 km behind a Wave Dragon farm (Wave Dragon 2009). Current onshore and attenuator devices have estimated energy capture ratios of ~50–93 per cent, though in practice existing devices may have less. A reduction in average power flux density may be less significant than the reduction in peak power flux densities, as results from the HEP test suggest that it is this that leads to many impacts. This is likely to be the case for wave and tidal energy. Terminator designs could be expected to have higher interception ratios and thereby the potential for higher energy extraction ratios could prove to be a significant factor in impacts. The number of conversions of energy form of the flow varies from three in the case of the Tapchan and OWC devices to one in the case of Pelamis.

It is suggested that the hypothesis might be usefully extended to water-based renewable energy sources and the environmental impacts appear to be related to the parameters proposed – power flux density, proportion of interception and energy capture, efficiency of energy conversion, and number of energy conversions – though no tests of data were made to confirm this.

9.2.4 Lower power flux density renewables

In Chapter 8 some of the parameters identified in the hypothesis were applied to the lower power flux density sources, below 1 kW m^{-2} average – wind, solar, biomass and geothermal – again on a theoretical basis, without extensive testing of data. Land use data of existing schemes appeared to indicate a potential relationship with power flux density in that lower power flux density sources resulted in greater land use, e.g. for solar and biomass production, while higher power flux densities conversely use less land, though the sample size was insufficient to confirm this statistically. However, as the variation in power flux densities may be less overall, e.g. solar energy fluxes vary by a factor of only ~2–2.5 (WEC 2004: 297) across the globe, the sensitivity of impacts to variations in power flux density may be less.

Low power flux density appears to result in greater compatibility with human habitation, for instance, solar PV involving only one direct energy conversion has few impacts and panels can be placed on roofs of buildings which are already occupying land. Solar PV then requires no extra land area. Similarly, biomass sources from waste streams can require little further land use too, since the land is already in use for productive purposes.

For low power flux sources, flow definition into discrete streams appears to be much poorer; air currents feature as broad flows of great height for instance, while solar radiation flows are 'field-like' in some respects due to reflection and refraction in the atmosphere. Biomass sources depend for their productivity on the biome characteristics of water and heat availability as well as light. Therefore, the notion of local flows has been proposed, so that the proportion of energy flux intercepted and extracted could then be determined locally.

This could apply to wind or central solar power stations or even biomass. For wind energy the low interception rate, together with the Betz limit and the need to disperse turbines, results in only a small proportion of the energy being extracted, unless the flow is of limited cross-section and volume. Solar energy conversion technologies currently achieve maximum efficiencies of ~30–40 per cent (Boyle 2004), and the proportion extractable is limited by this, although the proportion intercepted is often about ~33 per cent for central solar plants, whether PV or concentrating solar thermal type. Solar electric conversion efficiencies, though rising, rarely achieve 30 per cent and generally <15 per cent to 20 per cent is more likely for CSP or PV. For biomass, very low overall energy conversion efficiency is often found from solar to biomass energy end use units, e.g. about <0.5 per cent (Boyle 1996) to 2 per cent (Tan *et al.* 2009).

The number of energy conversions as a parameter of impact could be relevant for both solar and biomass, and wind too, though maybe to a lesser degree. For biomass, where a large number of conversions are carried out, this parameter also appears to be informative of impacts.

9.2.5 Land use compared to power flux density for the different sources

Examples of the onshore renewable energy sources' land use were compared with their power flux density for solar, biomass, wind and HEP. The power

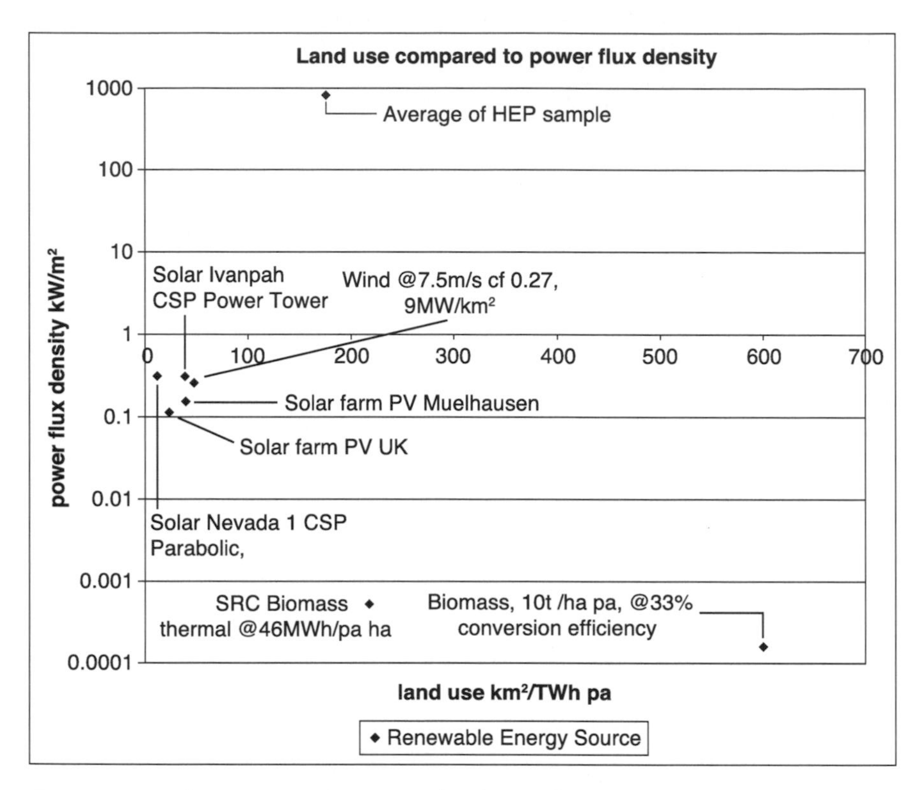

Figure 9.1 Land use compared to power flux density for some onshore renewable sources.

flux density for HEP is based on the head height. The results are shown in Figure 9.1.

The greatest land use was incurred by biomass production at 1,752 km^2/TWh p.a., which also had the lowest power flux density at 0.00016 kW m^{-2} average (based on 10 t ha^{-1} p.a. woody biomass and 33 per cent thermal conversion efficiency in the UK). The lowest land use was that of the Solar Thermal CSP Nevada 1, at 11.5 km^2/TWh p.a., followed by Three Gorges HEP scheme with 12.8 km^2/TWh p.a., and this scheme had the highest power flux density of 1,000.6 kW m^{-2}. However, when all the HEP schemes in the sample were included, the average land use was 175.58 km^2/TWh p.a., which was greater than that for wind energy at 46.97 km^2/TWh p.a. (based on 9 MW km^{-2}, 7.5 m s^{-1} wind speeds and a c.f. of 0.27). Solar energy required 38.87 km^2/TWh p.a., or as little as 11.5 km^2/TWh p.a. depending on latitude, resource and technology. For higher-latitude PV installation, the average power flux density was 154 W m^{-2}, based on estimations for Mühlhausen solar PV scheme in Germany, and 114 W m^{-2} for UK estimations, while for the lower-latitude desert CSP thermal plant it was 313 W m^{-2} (based on the Nevada 1 US site).

The results might indicate support for the hypothesis, in that higher power flux densities are generally associated with lower land use, but do not statistically confirm the hypothesis due to the size of the sample. The exception in the sample is the CSP plant with relatively low power flux density but low land use.

However, when the HEP schemes where the main purpose is water storage were removed from the sample (i.e. also those with high land use and low flow rates/high residence times), a somewhat clearer pattern could be observed. This is shown in Figure 9.2. If an equation were to be fitted to the data points in Figure 9.2, a power equation, $y = 2E+06x^{-3.8127}$ would fit best, with $R^2 = 0.6537$, or with the axes inverted (so that impact = power flux density × by factor) $y = 39.551x^{-0.1724}$.

This might indicate a correlation but only weakly. The use of land area for storage by HEP schemes might be compared with storage in biomass production, though this is an integral part of biomass chemical energy conversion processes, whereas for HEP the amount of storage can be

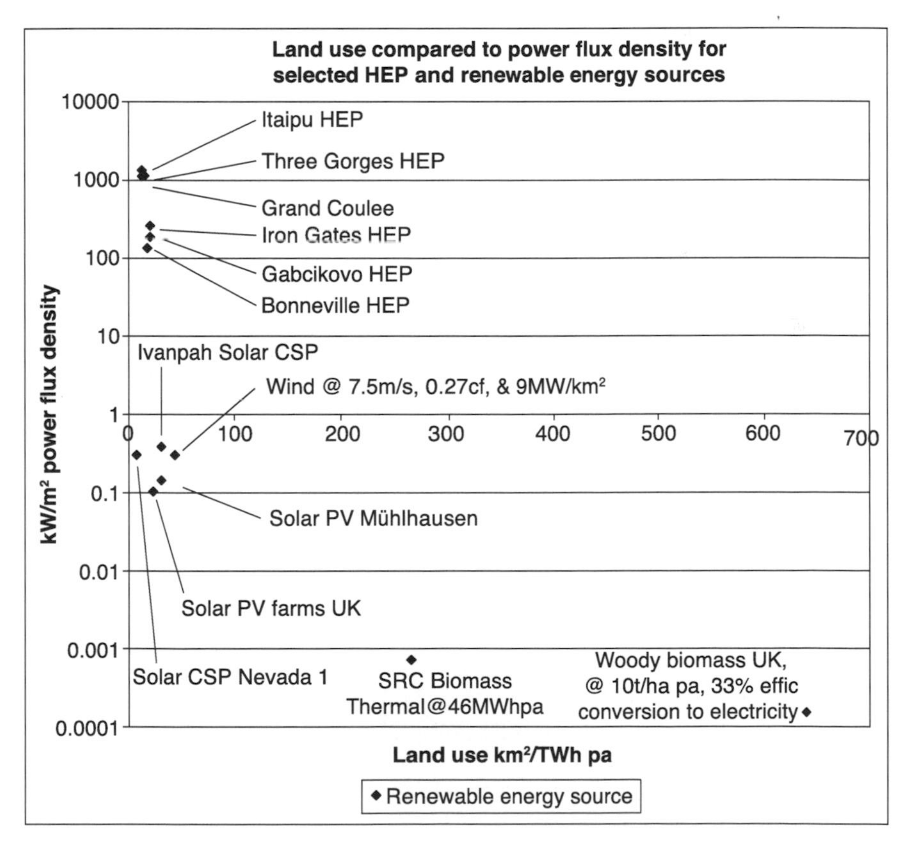

Figure 9.2 Land use compared to power flux density for selected HEP and renewable energy sources.

determined by design. The somewhat anomalous solar CSP result could be explained by the high average solar irradiance of 313 W m^{-2}, and also that almost all HEP schemes incorporate some storage involving land use. This HEP storage land use could theoretically be discounted, either by using water residence time in days / land use (km^2/TWh p.a.), or in practice by use of leats and zero storage. However, storage confers greater value for electricity generation.

The results might indicate a correlation between land use and power flux density for the onshore renewable energy sources, though the sample size was too small to be statistically representative.

9.3 Development and assessment of the hypothesis

The hypothesis began with impacts related to the parameter of power flux density, as reflected by the functions of natural energy flow and in land use. It was further developed to include the proportion or ratio of energy extracted from the flow, and interception of flow. The efficiency of conversion was an additional parameter, as reflected in unwanted by-products and also as a factor in the proportion of energy in the flow that can be extracted. The hypothesis proceeded to an additional parameter, that of conversion of the form of the energy in the flow and number of conversions, i.e. resulting in discontinuities in the functions.

Is the hypothesis confirmed?

The hypothesis is not statistically confirmed by the data obtained, since there are not enough data, but some of the results are indicative of support for the hypothesis. Overall, the hypothesis is not clearly contradicted by the data obtained. Of the various parameters tested on the reservoir reach, an inverse linear correlation between cross-sectional stream power and land use was indicated; there was an indication of a linear correlation between water residence time and land use; and a linear correlation between reservoir volume and land use was indicated. However, no correlation was found between land use and the parameters of power flux density derived from head height, former mean gradient, average reservoir fluid velocity, production factor and mean river flow. A correlation between reservoir sedimentation and loss of peak TSP below the dam was indicated, though not statistically confirmed.

For the stream power test downstream of the HEP schemes, data were indicative of support for the hypothesis, but not sufficient to provide statistically significant support for the hypothesis. A correlation between reduction in TSP post-HEP schemes, and sediment transport, reflected in delta land area loss, was indicated but not statistically supported due to the small sample size.

In conclusion, the study has not confirmed the hypothesis, but has shown some indication of support in certain respects, particularly stream power as a reflection of power flux density. When comparing power flux density (pfd) and land use for all of the renewable sources, for land use $I = 46.237\ (\mathrm{pfd})^{-0.1873}$, $R^2 = 0.736$, there could be an indication of support for the hypothesis, though again not statistically significant. In general the higher power flux density sources required less land, as might be expected; however, the solar CSP Nevada 1 could be somewhat anomalous, having relatively low power flux density, but also low land use requirement.

Further complexities

There are, however, additional complications associated with horizontal fluid flows, which are addressed by the second and further parts of the hypothesis, concerning conversions and efficiency. The proportion of the energy extracted from the flow might be a factor in impacts resulting from loss of the energy functions, such as sediment transport, though this was not tested.

Complexities are also introduced by the multiple purposes and uses of dams and reservoirs. Energy generation is rarely the sole purpose of a dam scheme. Flood control practices, navigation or water supply may result in modes of operation that cause or reduce impact. This requires careful analysis together with some assumptions as to the impact that can be allocated to the energy component.

Gaps

This study investigated one main aspect of the hypothesis, that of the power flux density in relation to the mechanical energy function of sediment transport, but has only tested this for selected HEP schemes, for the reservoir and then the river downstream, where data were available. Further mathematical testing of the significance of power flux density on a larger representative sample of HEP schemes would be required to confirm its significance. The additional parameters proposed – proportion, efficiency of conversion and number of conversions – could also be tested on a larger sample. The test only considered sediment transport and land use, but other impacts such as those on water quality and flora and fauna could be carried out.

The test of stream power changes to the river below the HEP scheme, which, due to the magnitude of the task, were only carried out on four rivers and for three deltas, could be extended to a larger sample. The land area loss of the Nile delta and other deltas that was estimated could be studied using satellite photographs.

More precise definitions of the relationship could be sought. For instance, do higher power flux densities in general result in more functions being performed? A precise description of the process of sediment erosion and transport and the relationship of water flow to sediment movement might have been incorporated into a larger study.

A computer model might have been constructed to model the relationship between the ratio of energy extracted from a natural energy flow and the impacts. This study has, though, largely been one exploring the parameters and dimensions of the concept. Given the complexity of the concept and specific nature of individual river systems and their sensitivity, the relationship between energy extraction and impacts is not likely to be a simple one, and has been outside the scope of this study.

9.4 Main conclusions

It can be concluded, in general, that power flux density could be a relevant parameter for determining aspects of impact. It is in general likely that, as might be expected, the greater the power flux density, the lower the land use or collector area required, once storage area is taken into account. Land use is a prime impact of renewable energy sources.

Power flux density changes could be a relevant parameter in determining impact from changes caused by the interception of fluxes of high power density, and the extraction of energy from them. This could tend to reduce geomorphic processes such as sediment removal and transport.

In the case of HEP, it is suggested that extraction of energy from the river tends to reduce energy available for sediment transport, but more importantly, perhaps, that impacts result from the change in peak energy flow of the river as a continuum, first in the reservoir itself and then where the flood is attenuated, all the way down the river to the mouth. In particular, the reduction in power flux density through attenuation of the flood flows TSP of the river appears to have the greatest effect in reduction of sediment transport, a prime impact.

Central statement

In bald terms, what emerges from this study is that the power flux density appears to relate to some impacts in the case of HEP and possibly other renewable sources, both in land use and in reduction of peak power flux density. However, the impact on land use relates inversely to the power flux density while for other types of impacts the relationship appears to be more proportional to changes in the power flux density.The ratio of energy extraction to the energy that would normally be used to perform natural functions, and to the total energy in the flow, may relate to the environmental impact.

9.5 Further research

The scope of this study has not extended to answering all the questions the hypothesis might raise. Here, the main thrust has been to test for evidence for the hypothesis, and investigate and test for validation of the parameters employed. If the relationship between power flux density and impacts can be fully confirmed, there will be many areas to which it can be applied, which will need to be the subject of further research work. For the parameters to be fully validated, tests would need to be carried out on a fully statistically representative sample of HEP schemes, as opposed to the limited sample, used in this study, as well as other renewable sources. However, collecting the extensive data required for this study entailed considerable difficulties, as much of the information is not readily available or even willingly released. It is recommended therefore that a database of HEP data and their impacts be set up.

Some of the further questions that can be posed with regard to HEP are provided below.

9.5.1 Further specific HEP questions

The question of whether rivers that have a greater power flux density are responsible for more mechanical erosion and deposition type functions can be posed. This question could be rephrased as 'does siting of HEP schemes on river reaches with steeper gradients lead to more or fewer impacts?' Concerning erosion, transport and deposition, greater flow rates, i.e. kinetic energy of moving water, are the consequence of either (1) steeper gradients, or (2) greater cubic flow supply, and also lower friction from greater cross-sectional areas through increased efficiency of flow in a viscous fluid.

Couched in geomorphologic terms, what type of rivers are responsible for more erosion/transport? Further questions might be posed based on the functions of the river in nature, that is, sediment transport, oxygenation, nutrient transport, organism transport and flushing.

The topic of sediment transport could be divided into stream load, bed load and dissolved in solution. Deposition or erosion resulting from the change in water flow rate, and TSP downstream of HEP schemes, could be further investigated.

Questions concerning water quality and the conditions for oxygenation could be posed, for instance what velocities and pressure differences are involved. The loss of 'white water' conditions resulting from inundation of falls and rapids from HEP plant could be compared to power flux density reductions, in addition to super saturation resulting from high pressures at some HEP schemes.

Other effects of flow rate changes – such as to nutrient transport, both inorganic and organic, organism transport, microscopic to plant and animal – could be investigated in terms of power flux density changes. Flushing effects of flow rates and waste product removal could form part of this. The PHABSIM suite of programmes relate flow regime and TSP to biological impact through habitat loss, which is an important consideration. Use of PHABSIM has become mandatory practice in the USA and increasingly in

Europe and elsewhere (USGS 2001). Further work in this area regarding HEP is recommended.

Extension of the hypothesis

The extension of the hypothesis in Chapter 7 to other water-based renewable sources – tidal barrage, marine current and wave energy converter – suggests the potential relevance of the hypothesis. When tidal barrages with higher power flux densities are compared with marine current turbines, impacts appear likely to be lower for marine current turbines. For wave energy the hypothesis appears relevant, higher power flux densities having potentially more functions, e.g. water mixing, bottom sediment movement and oxygenation, as well as shoreline erosion. Extraction of energy, reducing power flux density, could reduce these functions. Efficiency of conversion affects the energy extraction ratio, and the number of conversions that can be applied effectively to some devices, e.g. the Tapchan wave energy converter.

Extension of the hypothesis in Chapter 8 to the low power flux density sources, wind, solar and biomass, appears to indicate potential relevance of the hypothesis in terms of power flux density, proportion and conversion efficiency, to an extent. However, due to the low power flux densities in the range of less than 1 kW m^{-2} (~77 W m^{-2} to 10–12 kW m^{-2} for wind and ~0.1–1 kW m^{-2} for solar, and ~16 W m^{-2} to 320 W m^{-2} for biomass), differences in impacts, apart from land use, may be less apparent. The hypothesis application is complicated by the less defined nature of the flow more akin to a 'field'. Power flux density can be conceived of as a descriptor here of the flow definition, i.e. its cross-sectional boundaries.

Further research on other water based renewable energy sources

If the link between environmental impacts and power flux density proves to be true for rivers, does it also hold for other water-based flows, for instance wave, tidal, ocean current and other fluid flows such as wind, and other flows like solar and biomass? Or does it only apply to water flows? Further investigation of how such flows differ from those of rivers would be required. The interruption and interception of energy flows and changes in the power flux density involved in energy extraction might produce a reduction in the functions overall, or fewer functions being performed in total, or possibly different functions. Statistical testing of the loss of functions over significant periods would be required to link the environmental impact to change in functions caused by energy changes.

Further research on low power flux density sources

The application of the hypothesis and its parameters to the other low renewable energy sources such as wind, solar, biomass and geothermal could be explored further in terms of land use, power flux density, interception ratios, efficiency and number of conversions. Statistical tests of the parameters together with impacts would need to be carried out on representative samples.

Further possible impact parameters were proposed, but have proved to be beyond the scope and resources available in this study.

9.6 Concluding points

If it is accepted that the environmental impact from renewable sources may be related to power flux density, the implication is that interception and perturbation of a natural energy flow should not attenuate or reduce its peak power flux densities to any significant extent, since this is when the 'functions' of the flow are largely being performed. That is to say that river flood events perform functions and should be perpetuated, as also should high tidal ranges and flows in estuaries, as well as storm winter waves on beaches. In the design of novel renewable energy developments, the interruption of peak power flux densities should be avoided where high interception ratios and storage of a flow occur. In particular, this applies to HEP and tidal barrage technology, while less so to marine current turbines and wave energy converters.

For the low power flux density sources such as wind and solar, the flow is not intercepted or its form changed to the same degree, and the link with 'functions' in nature is possibly less likely to be affected. For instance, the power of storm force winds above 15 m s^{-1} is not harnessed by the turbine, but is instead 'spilled' as surplus; most of the energy harnessed would be derived from that part of the wind speed distribution with the most frequent speeds with sufficient power, and not the infrequent storm force winds. Therefore the device does not affect the higher power flux density wind speeds, which can still serve functions in nature. Similarly, this could apply to wave energy. It then becomes apparent that it is those sources where a large ratio of the flow can be intercepted and stored that appear to be prone to impacts derived from attenuating peak power flux densities; HEP, tidal barrages in particular, but also biomass and conceivably geothermal. The latter two cases involve further complexities which would require further consideration.

For HEP in particular, the implication is that 'run of the river' schemes without storage have considerably lower impacts than impoundment dams. This conclusion is not new, but the application of the concept of TSP modelling to HEP schemes is novel and appears to indicate support for the conclusion. If this conclusion is accepted, even on a per-kWh basis, and 'on line' flow storage is to be avoided (unless of course releases can simulate natural releases), then the increased variability of output from schemes without storage would have to be managed. Variability, it seems, is good for the environment, if a problem for reliable generation.

The implication of the hypothesis for all of the different renewable energy sources is that the majority of renewable energy should be harnessed from flows in or near their primary form – harnessing lower power flux density sources such as solar energy, before conversion to higher power flux densities, is likely to lead to lower overall environmental impact. This is somewhat counterintuitive to an engineering approach, where higher power density is preferable. If the hypothesis is correct, then to reduce environmental impact the majority of renewable energy extraction should be from solar energy. This energy extraction should only constitute a small proportion of the total flow. The number of energy conversions both in the natural flow and after capture should be minimised. Additionally, changes in the power flux density of the natural flow should be kept to a minimum. Where other sources are used for energy extraction, the same considerations apply. If energy is abstracted from dense high power flux density flows, the natural flow variability should be preserved, especially the peak energy flows, e.g. river floods, highest tides and waves. The point that the earth's natural energy systems appear to generate thermodynamic disequilibrium, though constantly rebalancing, in order to maximise the stages of entropic efficiency in energy conversion, can be perceived as the vital variability in most natural energy flows.

Bibliography

Boyle, G. (ed.) (1996) *Renewable Energy Power for a Sustainable Future*, Oxford University Press in conjunction with the Open University, Milton Keynes.

Boyle, G. (ed.) (2004) *Renewable Energy: Power for a Sustainable Future*, 2nd edition, Oxford University Press with the Open University, Oxford and Milton Keynes.

Charlier, R.H. (2003) 'Sustainable Co-Generation from the Tides: A Review', *Renewable and Sustainable Energy Reviews*, Vol. 7: 187–213.

Petts, G.E. (1984) *Ecology of Impounded Rivers: Perspectives for Ecological Management*, John Wiley & Sons, Chichester.

Tan, K.T., Lee, K.T., Mohamed, A.R. and Batia, S. (2009) 'Palm Oil: Addressing Issues and Towards Sustainable Development', *Renewable & Sustainable Energy Reviews*, Vol. 13: 420–7.

USGS (2001) 'Introduction to the Physical Habitat Simulation System', US Geological Survey, online resources: www.fort.usgs.gov/products/Publications/15000/chapter1.html, accessed 13 August 2010.

Wave Dragon (2009) 'Wave Dragon: Environment', online resources: www.wavedragon.net/index.php?option=com_content&task=view&id=5&Itemid=6, accessed 10 June 2009.

WEC (2004) *Comparison of Energy Systems Using Life Cycle Assessment*, World Energy Council, London.

Index